Anonymous

Freight Charges for Ocean Transportation of the Products of

Agriculture, October 1, 1895,

to October 1, 1896

Anonymous

Freight Charges for Ocean Transportation of the Products of Agriculture, October 1, 1895,
to October 1, 1896

ISBN/EAN: 9783337318475

Printed in Europe, USA, Canada, Australia, Japan

Cover: Foto ©berggeist007 / pixelio.de

More available books at **www.hansebooks.com**

MISCELLANEOUS SERIES, BULLETIN No. 12.

U. S. DEPARTMENT OF AGRICULTURE.

DIVISION OF STATISTICS.

FREIGHT CHARGES

FOR

Ocean Transportation of the Products of Agriculture.

OCTOBER 1, 1895, TO OCTOBER 1, 1896.

WASHINGTON:

GOVERNMENT PRINTING OFFICE.

1896.

LETTER OF TRANSMITTAL.

U. S. DEPARTMENT OF AGRICULTURE,
DIVISION OF STATISTICS,
Washington, D. C., October 17, 1896.

SIR: The accompanying report on "Freight charges for ocean transportation of the products of agriculture" is respectfully submitted, with the recommendation that it be printed as Bulletin No. 12 (Miscellaneous Series) of this Division.

All of the rates shown have been furnished by officials of the lines over which they apply, and the thanks of the Department are due them for their kind compliance with its requests.

Respectfully,

HENRY A. ROBINSON.
Statistician.

Hon. J. STERLING MORTON,
Secretary.

CONTENTS.

	Page.
Introductory	7
Value of domestic exports	7
Transatlantic traffic rates	9
Coastwise traffic rates	35
Freight charges in England on agricultural products	43
The complaints of the English farmer	43
The efforts of the railroads	43
The London and Northwestern Railway Company	44
The Great Western Railway Company	44
The Great Eastern Railway Company	48
The Southeastern Railway Company	49
Typical English rates on agricultural products	51
London and Northwestern Railway rates by merchandise train (ordinary and express) to London	51

FREIGHT CHARGES FOR OCEAN TRANSPORTATION OF THE PRODUCTS OF AGRICULTURE.

INTRODUCTORY.

VALUE OF DOMESTIC EXPORTS.

The total value of articles of merchandise of domestic origin, exclusive of gold and silver, exported from the United States during the year ending June 30, 1895, was $793,392,599, of which $553,210,026, or nearly 70 per cent, represents the value of exported products of farms. In the following pages an attempt is made to show the freight charges for transporting the principal of these products from American to European ports, and a few tables are added showing similar charges for coastwise transportation between ports located within the United States. There has also been added an article and several tables prepared by Mr. Lorin A. Lathrop, United States consul at Bristol, Great Britain, explaining some recent developments of the policy of the principal English railways toward the agricultural interests of that country.

The table first following shows, for each article for which freight charges appear in the subsequent pages, the value of the total quantity of domestic origin exported during the fiscal year 1894-95, and the percentage of that total exported, respectively, from Baltimore, Boston, New York, and all other ports. The aggregate value of domestic exports of the articles shown in this statement during the year referred to was $516,558,109, or 93 per cent of the total value of all exports of farm products during that year, while of this total 55.46 per cent was exported from the three ports of Baltimore, Boston, and New York.

Table showing the value of domestic exports during the year ending June 30, 1895, and the proportions, respectively, exported from Baltimore, Boston, New York, and all other ports.

Articles.	Value of domestic exports.	Per cent from—			
		Baltimore.	Boston.	New York.	All other ports.
Cattle	$30,603,796	14.98	35.88	38.23	10.91
Hogs	72,424	.79		3.47	95.74
Horses	2,209,298	3.30	2.59	51.85	42.26
Mules	186,452	5.19		36.27	58.54
Sheep	2,630,686	23.17	33.52	33.75	9.56
Barley	767,228		2.83	1.97	95.20
Corn	14,650,767	17.10	8.88	28.84	45.18
Corn meal	648,844	7.18	6.82	77.48	8.52
Oats	200,793	.15	.31	40.79	58.75
Oatmeal	566,321	18.40	20.87	48.27	12.46
Rye	5,340			90.37	9.63
Wheat	43,805,663	10.29	6.86	30.05	52.80
Wheat flour	51,651,928	16.49	14.32	36.80	32.40

7

Table showing the value of domestic exports during the year ending June 30, 1895, etc.—Con.

Articles.	Value of domestic exports.	Per cent from—			
		Baltimore.	Boston.	New York.	All other ports.
Cotton, unmanufactured	$204,900,990	4.06	4.64	14.69	76.61
Apples, dried	461,214	6.70	.04	92.50	.76
Apples, green or ripe	1,054,318	.01	55.88	32.81	11.30
Hay	699,029	.07	28.95	51.71	19.27
Hops	1,872,597	....	3.67	90.26	6.07
Cotton-seed oil	6,813,313	6.98	.27	55.94	36.81
Beef, canned	5,720,933	34.17	4.45	47.05	14.33
Beef, fresh	16,832,860	2.45	45.20	49.30	3.05
Beef, salted or pickled	3,558,230	20.64	7.74	59.20	12.42
Tallow	1,293,059	40.59	3.34	38.34	17.73
Bacon	37,776,293	3.75	37.70	49.28	9.27
Hams	10,960,567	2.15	50.41	35.75	11.69
Pork, pickled	4,138,400	5.79	8.33	70.64	15.24
Lard	36,821,508	16.26	17.60	59.83	6.25
Butter	915,533	.11	8.69	76.68	14.52
Cheese	5,497,539	....	7.46	73.67	18.87
Clover seed	2,124,997	32.04	.62	51.70	15.64
Tobacco, unmanufactured	25,798,968	20.71	1.23	52.46	25.60
Potatoes	418,221	.01	....	76.59	23.40
Total	516,558,109	9.16	13.61	32.69	44.54

The following table shows for the articles enumerated in the preceding table the percentages of the value of domestic exports destined to the United Kingdom of Great Britain and Ireland, to other European countries, and to all other countries of the world. These introductory tables are believed to afford a basis for estimating the importance of the tables of rates which immediately follow them.

Table showing the value and distribution of exports of certain articles of domestic origin during the year ending June 30, 1895.

Articles.	Value of domestic exports.	Per cent shipped to—		
		United Kingdom.	Other European countries.	All other countries.
Cattle	$30,603,796	93.10	6.11	0.79
Hogs	72,424	.75	....	99.25
Horses	2,209,298	43.12	17.30	39.58
Mules	186,452	....	....	100.00
Sheep	2,630,686	89.29	2.67	8.04
Barley	767,228	80.96	.12	18.92
Corn	14,650,767	55.47	28.01	16.52
Corn meal	648,844	27.80	2.10	70.10
Oats	200,793	1.36	3.03	95.61
Oatmeal	566,321	73.96	25.21	.83
Rye	5,340	89.89	....	10.11
Wheat	43,805,663	69.52	24.25	6.23
Wheat flour	51,651,928	59.11	8.40	32.49
Cotton, unmanufactured	204,900,990	50.81	45.98	3.21
Apples, dried	461,214	20.44	77.03	2.53
Apples, green or ripe	1,054,318	94.07	1.12	4.81
Hay	699,029	67.47	4.84	27.69
Hops	1,872,597	95.21	.14	4.65
Cotton-seed oil	6,813,313	13.60	70.67	15.73
Beef, canned	5,720,933	62.28	24.48	13.24
Beef, fresh	16,832,860	99.72	.19	.09
Beef, salted or pickled	3,558,230	46.19	24.68	29.13
Tallow	1,293,059	17.63	50.63	31.74
Bacon	37,776,293	76.83	15.35	7.82
Hams	10,960,567	84.35	6.04	9.61
Pork, pickled	4,138,400	24.95	5.47	69.58
Lard	36,821,508	38.84	45.40	15.76
Butter	915,533	14.74	6.61	78.65
Cheese	5,497,539	79.36	.11	20.53
Clover seed	2,124,997	46.87	38.22	14.91
Tobacco, unmanufactured	25,798,968	36.03	55.69	8.28
Potatoes	418,221	.44	....	99.56
Total	516,558,109	58.46	31.48	10.06

TRANSATLANTIC TRAFFIC RATES.

The following tables show the traffic rates between different ports by the various transportation companies for the named commodities exported to European countries from October 1, 1895, to October 1, 1896, by months:

NEW YORK TO SOUTHAMPTON VIA AMERICAN LINE.

Commodities.		1895 Oct. 1.	Nov. 1.	Dec. 1.	Jan. 1.	Feb. 1.	Mar. 1.	Apr. 1.	May 1.	June 1.	July 1.	Aug. 1.	1896 Sept. 1.
Wheat	per bushel	$0.07½	$0.08½	$0.08½	$0.04½	$0.08½	$0.08½	$0.06½	$0.08½	$0.06½	$0.06½	$0.06½	$0.08½
Corn	do.	.07½	.08½	.08½	.04½	.08½	.08½	.06½	.08½	.06½	.06½	.06½	.08½
Flour: In barrels	per barrel	.487	.487	.487	.487	.487	.426	.365	.426	.426	.426	.487	.547
In sacks	per 2,240 lbs	2.43	3.04	3.04	3.04	3.65	2.73	2.43	2.73	2.73	3.04	3.04	3.65
Clover seed	do.	3.65	4.86	4.26	4.26	4.86	4.26	3.65	3.65	3.65	3.65	4.26	4.86
Bacon	do.	3.65	4.26	4.26	4.26	4.26	3.65	3.65	3.65	3.65	3.65	3.65	4.26
Hams	do.	3.65	4.26	4.26	4.26	4.26	3.65	3.65	3.65	3.65	3.65	3.65	4.26
Lard	do.	4.86	5.47	5.47	5.47	5.47	4.86	4.86	4.86	4.86	4.86	4.86	5.47
Lard, in small packages	do.	3.65	4.26	4.26	4.26	4.26	3.65	3.65	3.65	3.65	3.65	3.65	4.26
Tallow	per tierce	.73	.852	.852	.852	.852	.73	.73	.73	.487	.73	.73	.852
Beef	per barrel	.547	.608	.608	.608	.608	.487	.487	.487	.487	.447	.487	.608
Pork	per barrel	6.08	9.73	7.30	7.30	7.30	7.30	6.08	7.30	7.30	6.08	6.08	7.10
Butter	per 2,240 lbs	4.86	8.51	6.08	6.08	6.08	6.08	4.86	6.08	4.86	4.86	4.86	6.08
Cheese	do.	3.04	3.65	3.65	3.65	4.86	3.65	3.65	3.65	3.65	3.65	3.65	4.26
Cotton-seed meal	do.	.73	.852	.852	.852	.852	.71	.71	.71	.73	.71	.71	.852
Cotton-seed oil	per barrel	1.00	1.00	.75	1.00	.65	.75	.75	.75	.85	.75	.85	.85
Hops	per 100 lbs										.75	.85	.85
Lumber: Hard	per 2,240 lbs	4.46	4.26	4.86	4.86	4.46	3.65	3.65	3.65	4.26	4.26	4.26	4.86
Soft	do.	4.46	4.86	6.08	5.47	6.08	4.26	4.26	4.86	4.86	4.86	4.46	5.47
Oil cake	do.	2.43	3.04	3.04	3.65	3.65	1.825	2.07	2.43	2.73	3.04	3.04	3.65
Resin	per 280 lbs	.426	.487	.487	.547	.608	.365	.365	.426	.426	.487	.426	.487
Tobacco: In cases	per 40 cubic feet	3.65	4.86	4.26	4.86	4.86	3.65	3.65	4.26	3.65	4.26	4.86	4.86
In hogsheads	per hogshead	4.86	6.08	6.08	6.08	6.08	4.86	4.86	5.47	5.47	6.08	6.08	6.08
Apples: Green	per barrel	.73	.606	.73	.71	.71	.608	.608	.608	.852	.73	.73	.852
Dried	per 2,240 lbs	4.86	4.86	5.47	4.86	4.86	4.26	4.26	4.26	4.26	4.26	4.26	4.86
Hay, pressed in bales	do.	6.08	7.30	6.08	7.30	6.08	4.86	4.86	5.47	6.08	4.86	4.26	4.86
Measurement	per ton of 40 cubic feet	3.65	7.20	6.08	4.86	6.08	6.08	6.08	6.08	4.86	6.08	6.08	6.08
Primage	per cent	5	5	5	5	5	5	5	5	5	5	5	5

Transatlantic traffic rates—Continued.

NEW YORK TO LIVERPOOL, VIA WHITE STAR LINE.

Commodities			1895			1896								
			Oct. 1	Nov. 1	Dec. 1	Jan. 1	Feb. 1	Mar. 1	Apr. 1	May 1	June 1	July 1	Aug. 1	Sept. 1
Wheat	per bushel		$0.046	$0.07	$0.06	$0.065	$0.06	$0.04	$0.025	$0.03	$0.04	$0.01	$0.04	$0.065
Corn	do		.045	.07	.06	.065	.06	.01	.025	.01	.04	.04	.04	.065
Flour: In barrels	per barrel		.304	.48	.48	.48	.36	.36	.24	.24	.25	.24	.36	.48
In sacks	per 2,240 lbs		2.13	2.40	2.40	2.40	2.40	2.10	1.50	1.50	1.80	1.80	2.10	3.00
Clover seed	do		3.04	3.60	3.60	3.60	3.60	3.00	2.40	2.40	2.40	2.40	3.60	3.60
Bacon	do		2.13	3.60	3.60	3.60	3.00	2.40	2.40	1.80	2.40	1.80	2.40	3.00
Hams	do		2.43	4.80	3.60	3.60	3.60	2.40	2.40	1.80	2.40	1.80	2.40	3.00
Lard	do		2.13	3.00	3.60	3.60	2.70	2.40	2.40	1.80	2.40	1.80	2.40	3.00
Lard, in small packages	do		2.43	4.80	4.80	4.20	3.60	3.00	3.00	2.40	3.00	2.40	3.00	3.60
Tallow	do		2.13	3.60	3.60	3.60	3.60	2.40	1.80	1.80	2.40	1.80	2.40	3.00
Beef	per tierce		.487	.72	.66	.66	.66	.44	.48	.36	.36	.48	.48	.60
Pork	per barrel		.365	.48	.24	.48	.36	.36	.30	.30	.30	.36	.42	.48
Butter	per 2,240 lbs		6.08	9.60	7.20	7.20	6.00	6.00	6.00	6.00	7.20	6.04	7.20	7.20
Cheese	do		4.86	6.00	6.00	6.00	4.80	4.80	4.80	4.80	4.80	4.80	4.80	4.80
Cotton-seed meal	do		2.43	2.40	3.00	2.40	3.60	2.40	1.80	1.50	2.40	1.80	2.40	3.00
Cotton-seed oil	per barrel		.426	.60	.72	.72	.72	.54	.48	.48	.54	.48	.60	.60
Cotton	per lb		.00¾	.00¾	.00³⁄₁₆	.001	.00¼	.00¾	.00¾	.00¾	.00⅞	.00½	.00¼	.00¾
Hops	per 100 lbs		.50	.438	.375	.375	.22	.375	.375	.575	.575	.50	.50	.625
Lumber: Hard	per 2,240 lbs		2.43	2.40	3.00	3.60	3.00	3.00	2.40	2.40	2.40	2.40	2.40	2.40
Soft	do		3.04	3.60	3.60	4.20	3.60	3.60	3.00	3.00	3.00	3.00	3.60	3.60
Oil cake	do		2.13	2.40	2.40	3.40	2.40	1.80	1.50	1.50	1.80	1.80	2.40	3.00
Rosin	per 280 lbs		.27	.30	.48	.48	.36	.30	.18	.18	.30	.30	.36	.43
Tobacco: In cases	per 40 cubic feet		3.65	3.60	3.00	3.60	2.40	2.40	2.40	2.40	2.40	2.40	2.40	2.40
In hogsheads	per 100 lbs		.18	.25	.21	.20	.20	.20	.18	.18	.16	.15	.20	.35
Apples	per barrel		.487	.60	.36	.60	.60	.60	.60	.60			.60	.60
Green	per 2,240 lbs		3.65	3.60	3.00	3.60	3.60	3.40	3.00	3.00	3.60	2.40	3.60	4.80
Dried	do		4.86	6.00	4.80	7.20	7.20	6.00	4.80	2.40	3.60	3.60	4.80	9.60
Hay, pressed, in bales	per head		7.91	7.20	7.20	10.80	9.45	10.30	9.00	9.60	9.60	8.40	6.00	9.60
Cattle	do													
Sheep	do		.97	1.20	1.20	2.40	1.20	1.08	1.08	1.08	.96	.84	.84	1.00
Measurement	per ton of 40 cubic feet		2.41	2.40	2.40	2.40	2.40	2.40	2.40	2.40	2.40	2.40	2.40	2.00
				4.80	4.80	4.80	4.80	4.80	4.80	4.80	4.80	4.80	4.80	4.80
Primage	per cent		5	5	5	5	5	5	5	5	5	5	5	5

BOSTON TO LIVERPOOL VIA LEYLAND LINE.

		$0.036	$0.061	$0.061	$0.061	$0.061	$0.051	$0.02	$0.015	$0.03	$0.04	$0.036	$0.04	$0.04	$0.061	$0.061
Wheat	per bushel	.036	.061	.061	.061	.061	.051	.02	.015	.03	.04	.036	.04	.04	.061	.061
Corn	do	.086														
Rye	do	.051														
Oats	do	.051														
Barley	do	.051														
Flour:																
In barrels	per barrel	.365	.365	.365	.365	.365	.304	1.217	.973	1.217	1.825	1.825	1.825	2.13	.365	.426
In sacks	per 2,240 lbs	1.70	2.71	2.43	2.43	2.43	2.13	1.217	.973	1.217	1.825	2.41	2.13	3.04	2.43	3.04
Corn meal	do	1.70	2.71	2.43	2.43	2.43	2.43	1.217	.973	1.217	2.41	2.43	3.04	3.04	3.04	3.04
Oatmeal	do	1.70	2.71	2.43	2.43	2.43	2.43	1.217	.973	1.217	3.04	2.43	3.65	3.65	3.65	3.04
Clover seed	do	2.04	2.71	2.43	2.43	3.04	2.43	1.217	.852	1.217	3.01	2.43	3.04	3.04	3.04	3.04
Beef:																
Canned	do	1.70	3.65	3.65	3.74	3.74	3.65	1.217	3.65	3.65	1.825	1.46	1.825	3.04	3.04	3.04
Fresh	per 40 cubic feet	3.65	.73	.487	.487	.487	.73	1.825	.852	1.217	.487	1.825	.487	3.04	2.43	.487
In barrels	per barrel	.243	3.34	3.04	3.04	3.04	3.34	1.825	.85	1.825	1.825	2.43	1.825	3.65	1.825	3.04
Tallow	per 2,240 lbs	1.70	.487	.365	.365	.365	.487	1.217	.852	1.217	1.825	2.43	1.825	3.01	3.04	.487
Pork	per barrel	.243	3.65	3.14	3.14	3.04	3.65	1.825	1.46	1.825	1.825	1.46	1.825	9.73	8.51	3.04
Bacon	per 2,240 lbs	1.70	3.65	3.74	3.74	3.04	3.65	1.217	1.46	1.217	1.825	1.825	1.825	1.825	1.217	3.04
Hams	do	1.70				3.04		1.217	1.46	1.217	.10	.10	.10		1.825	3.04
Lard:																
In tierces	do	1.70	3.34	2.43	2.43	2.43	3.34	1.217	.852	1.217	1.825	1.825	1.825	2.13	2.43	3.65
In small packages	do	2.43	4.86	3.65	3.65	3.65	4.86	1.825	1.46	1.825	2.41	2.41	3.04	3.65	3.04	3.65
Butter	do	2.43	4.86	3.65	3.65	3.04	4.86	1.825	1.46	1.825	3.04	2.43	3.04	3.65	3.04	3.65
Cheese	do	2.43	4.86	3.65	3.65	4.26	4.86	1.825	1.46	1.825	3.01	2.13	3.01	3.65	3.65	3.04
Cotton-seed meal	do	2.13	2.92	2.43	2.41	2.43	2.92	1.217	.973	1.217	1.825	1.46	1.825	3.04	3.65	3.04
Cotton-seed oil	do	2.13	3.65	3.65	3.65	3.04	3.65	1.825		1.825	.10	1.825	.10	3.04	3.04	3.04
Cotton	per 100 lbs	.23	.20	.14	.14	.14	.20	.15	.08	.09	.10	.10	.10	.20	.20	.20
Hops	per 2,240 lbs	9.73												.20		
Lumber:																
Hard	do	2.13	3.04	3.04	3.04	3.04	3.04	1.825	1.58	1.825	2.13	2.13	2.13	2.43	3.04	3.04
Soft	do	3.04	3.71	3.65	3.65	3.65	3.71	1.217	.973	1.217	1.43	3.04	1.825	3.04	3.65	3.65
Oil cake	do	1.70	3.04	3.19	3.19	3.04	3.04	1.217		1.217	1.46	1.46	1.46	1.825	3.04	3.04
Rosin	do	1.70		3.04	3.04	3.04		1.217		3.65	1.46	1.46	1.46	1.825	3.04	3.04
Tobacco:																
In cases	per 40 cubic feet	1.825	2.43	1.825	1.825	2.43	2.43	1.825	1.58	1.825	1.825	1.825	1.825	1.825	1.825	1.825
In hogsheads	per 2,240 lbs	5.47	6.08	7.30	7.30	4.48	6.08	3.36	2.688	3.65	3.65	3.65	3.65	3.65	3.65	4.26
Apples:																
Green	per barrel	.365	.365	.365	.365	.365	.365	.304	.304	.365	.187	.365	.365	.365	.187	.365
Dried	per 40 cubic feet	1.825	1.825	1.825	1.825	1.825	1.825	1.825	1.46	1.825	1.825	1.825	1.825	1.825	1.825	1.825
Hay, pressed, in bales	per 2,240 lbs	6.08	7.30	8.51	8.51	6.08	7.30	9.71	9.73	9.71	3.04	3.04	3.01	9.73	3.04	3.05
Cattle	per head	6.08	7.30	9.73	9.73	10.95	7.30	9.73		9.71	6.08	8.51	9.73		6.08	8.51
Sheep	do	.973	1.217	1.217	1.217	1.217	1.217	.973	.973	.973	1.217	.971	9.73	9.71	1.217	1.217
Measurement	per ton of 40 cubic feet	1.825	1.43	1.825	1.825	1.825	1.43	1.825	1.16	1.825	1.825	1.825	1.825	1.825	1.825	1.825
Primage	per cent	5	5	5	5	5	5	5	5	5	5	5	5	5	5	5

Transatlantic traffic rates—Continued.

NEW YORK TO LONDON VIA WILSON LINE.

Commodities.		1895.						1896.					
		Oct. 1.	Nov. 1.	Dec. 1.	Jan. 1.	Feb. 1.	Mar. 1.	Apr. 1.	May 1.	June 1.	July 1.	Aug. 1.	Sept. 1.
Wheat	per 60 lbs	$0.061	$0.066	$0.076	$0.076		$0.025 .03	$0.036	$0.051	$0.061	$0.061	$0.066	$0.071
Corn	do	.061	.066	.076	.076		.025 .03	.036	.051	.061	.061	.066	.071
Rye	do	.061	.066	.076	.076		.025 .03	.036	.051	.061	.061	.066	.071
Oats	per 320 lbs	.365	.426	.608	.426		.487		.426	.126	.487	.487	.487
Barley	per 48 lbs	.061	.061	.068	.076		.025 .03	.036	.051	.061	.061	.066	.071
Flour:													
In barrels	per barrel	.365	.487	.187	.365		.243	.243	.243	.365	.126	.487	.487
In sacks	per ton	2.43	3.04	3.04	3.04		1.52	1.58	1.82	2.43	2.43	3.04	3.04
Corn meal, in sacks	do	2.43	3.04	3.04	3.65		1.52	1.58	1.82	2.43	2.43		
Oatmeal, in sacks	do	2.43	3.04	3.04	3.04		1.52	1.58	1.82	2.43	2.43		a. 487
Clover seed, in sacks	do	3.65	3.65	4.26	4.26		3.04	3.04	3.04	3.65	3.43	3.65	3.65
Beef:													
Canned	do	3.65	3.65	4.26	4.86		3.34	3.43	3.43	3.43	3.04	3.65	3.65
In tierces	per tierce	.608	.73	.852	.852		.608	.547	.487	.487	.608	.608	.669
Tallow	per ton	3.01	3.65	3.65	4.26		3.04	3.43	3.43	3.43	3.04	3.04	3.04
Pork	per barrel	.426	.487	.608	.608		.487	.365	.365	.365	.487	.487	.487
Bacon	per ton	3.04	3.65	4.26	4.26		3.04	3.43	3.43	3.43	3.04	3.04	3.04
Hams	do	3.04	3.65	4.26	4.26		3.01	3.43	3.43	3.43	3.04	3.04	3.34
Lard:													
In tierces	do	3.04	3.65	4.26	4.26		3.04	3.43	3.43	3.43	3.04	3.04	3.34
In small packages	do	3.05	4.26	4.86	4.86		3.65	3.04	3.04	3.04	3.65	3.65	3.95
Butter	do	6.08	7.30	6.08	6.08		6.08	4.86	6.08	6.08	6.08	6.08	6.08
Cheese	do	4.86	4.86	4.86	4.86		4.86	3.65	4.86	4.86	4.86	4.86	4.86
Cotton-seed meal	do	3.65	3.65	3.65	3.65								
Cotton-seed oil	per barrel	.608	.73	.852	.973		.547	.547	.487	.608	.608	.669	.73
Hops	per 100 lbs	.459	.459	.459	.438		.459	.459	.459	.55	.608		
Lumber:													
Hard	per ton	3.65	3.65	3.65	4.26		3.04	3.04	3.43	3.04	3.04	3.65	3.65
Soft	do	4.26	4.26	4.26	4.86		3.65	3.65	3.04	3.65	3.65	4.26	4.26
Oil cake	do	2.73	2.92	3.04	3.04		1.52	1.52	1.82	2.43	2.43	3.04	2.73
Rosin	per 280 lbs	.365	.487	.487	.608		.182	.243	.243	.487	.487	.487	.487
Tobacco:													
In cases	per 40 cubic feet	3.65	4.86	4.26	4.86		2.43	2.43	2.43	3.04	3.04	3.65	
In hogsheads	per 100 lbs	.18	.20	.55	.20		.18	.18	.18	.18	.20	.22	.22

Apples:												
Green	per barrel	.487	.608	.608	.608	.608	.487	.487	.487	.487		.608
Dried	do	.487	.608	.608	.608	.608	.487	.487	.487	.487		.608
Hay, pressed in bales	per ton	3.65	3.05	7.30	7.30	6.08	3.65	3.65	3.65	3.65		.73
Cattle	per head	7.30	6.08	8.51	8.51	10.95	9.73	9.73	9.73	9.73	8.51	7.30	9.73
Sheep	do	.852	.973	.973	.973	.973	.973	.973	.973	.973	.973	.971	.973
Measurement	per ton of 40 cubic feet	2.43	3.04	4.26	4.26	3.65	2.43	2.43	2.43	2.43	3.04	3.04	3.04
Primage	per cent	3.65	3.65				3.65	3.65	3.65	3.65			

BOSTON TO LONDON VIA JOHNSTON LINE.

Wheat	per bushel	$0.051	$0.066	$0.066	$0.071	$0.066	$0.066	$0.041	$0.036		$0.061	$0.001	$0.061	$0.076
Corn	do	.051	.066	.066	.071	.066	.066	.041	.036		.061	.071	.061	.076
Rye	do	.051	.071	.071	.076	.071	.071	.365	.243		.071	.071	.061	.076
Oats	per quarter	.913	.608	.973	.973	.426	.426	.365	.243		.426	.126	.487	.487
Barley	do	.73	.608	.973	.973	.426	.426	.365	.243		.426	.426	.487	.487
Flour:														
In barrels	per ton	3.65	3.65	3.65	3.65	3.65	2.43	2.13	1.70		2.43	3.04	3.34	3.65
In sacks	do	3.04	3.34	3.34	3.04	3.34	2.43	2.13	1.94		2.13	2.31	2.73	3.04
Corn meal	do	3.34	3.65	3.34	3.34	3.34	2.67	2.43	1.94		2.43	2.43	3.04	3.34
Oatmeal	do	3.34	3.65	3.34	3.34	3.04	2.04	2.79	2.31		2.13	2.43	3.04	3.34
Clover seed	do	3.65	8.51	3.65	3.65	3.65	2.73	2.55	2.07		3.65	3.65	3.65	3.65
Beef:														
Canned	do	3.04	3.65	4.26	3.34	3.65	3.65	2.43	2.19		2.43	3.65	3.04	3.65
Fresh	per 40 cubic feet		3.65	4.26			a.608	a.487	a.365					
In barrels	per ton	3.65	4.26	4.26	4.26	4.26	3.65	2.73	2.19		3.65	3.65	3.65	3.65
Tallow	do	3.65	4.26	4.26	4.26	4.26	3.65	2.73	2.19		3.65	3.65	3.65	3.65
Pork	do	3.65	4.26	4.26	4.26	4.26	3.65	2.73	2.19		3.65	3.65	3.65	3.65
Bacon	do	3.65	4.26	4.26	4.26	4.26	3.65	2.73	2.19		3.65	3.65	3.65	3.65
Hams	do	3.65	4.26	4.26	4.26	4.26	3.65	2.73	2.19		3.65	3.65	3.65	3.65
Lard:														
In tierces	do	3.65	3.65	4.26	4.26	4.26	3.65	2.73	2.19		3.65	3.65	3.65	3.65
In small packages	do	4.26	4.20	4.80	4.86	4.56	4.26	3.34	2.73		4.26	4.26	4.26	4.26
Butter	do	4.56	4.86	4.26	4.26	6.08	6.08	6.08	4.86		4.26	4.86	4.86	4.86
Cheese	do	4.26	4.86	4.26	4.26	4.26	4.46	4.26	4.26		4.26	4.86	4.86	4.86

a Per barrel.

Transatlantic traffic rates—Continued.

BOSTON TO LONDON VIA JOHNSTON LINE—Continued.

Commodities		1895			1896								
		Oct. 1	Nov. 1	Dec. 1	Jan. 1	Feb. 1	Mar. 1	Apr. 1	May 1	June 1	July 1	Aug. 1	Sept. 1
Cotton-seed meal	per ton	$3.34	$4.86	$3.65	$3.65					$4.26	$2.43	$3.65	$3.65
Cotton-seed oil	do	3.65	1.217	3.65	3.65				$0.487	.487	3.05	3.65	3.65
Cotton	per lb.	.056	.071	.008			$0.608	$0.365	$0.487				
Hops	per ton	16.80	10.95	14.00	9.117	$14.56	12.77	12.77	12.77	12.32	14.56	14.56	16.80
Lumber:													
Hard	do	3.65	3.04	3.34	3.34	3.65	3.01	2.43	3.04	3.34	3.04	3.34	3.04
Soft	do	4.56	3.95	4.26	4.26	4.86	1.95	3.28	3.95	4.26	4.26	4.26	4.26
Oil cake	do	3.04	3.34		3.04	3.04	2.43	1.70	2.13	2.13	2.31	2.73	3.04
Rosin	per barrel	a3.04	a3.65		.608	1.217	.304	.243	.304		1.217	.973	.73
Tobacco:													
In cases	per ton	6.08	6.08	6.08	6.08	6.08	6.08	6.08	6.08	6.08	6.08	6.08	4.86
In hogsheads	do	6.08	6.08	6.08	6.08	6.08	6.08	6.08	6.08	6.08	6.08	6.08	4.86
Apples:													
Green	per barrel	.487	.608	.487	.487	.426	.608	.487	.608		.487	.73	.608
Dried	per ton	4.86	6.08	3.04	3.65	4.26	4.26	3.65	4.26	2.43	3.65	3.65	3.65
Hay, pressed, in bales	do	4.86	4.86	7.30	10.95	7.30	7.30	7.30	7.30	8.51	8.51	8.51	8.51
Cattle	per head	7.30	6.08	8.51		12.16	12.16	8.51	8.51	9.73	10.34		10.95
Sheep	do	1.217	1.095	.973	1.46	1.46	1.46	1.217	1.34	1.217	1.46	1.217	1.947
Measurement	per ton of 40 cubic feet	3.65	3.65	3.65	3.65	3.65	3.65	3.65	3.65	2.43	3.65	3.65	3.65
Primage	per cent.	5	5	5	5	5	5	5	5	5	5	5	5

NEW YORK TO GLASGOW VIA ANCHOR LINE.

Commodities		Oct. 1	Nov. 1	Dec. 1	Jan. 1	Feb. 1	Mar. 1	Apr. 1	May 1	June 1	July 1	Aug. 1	Sept. 1
Wheat	per bushel	$0.056	$0.071	$0.071	$0.071	$0.061	$0.041	$0.005	$0.02	$0.046	$0.011	$0.051	$0.076
		.061	.071	.076				.015	0.03				.081
Corn	do	.056	.061	.071	.071	.061	.041	.065	.02	.046	.041	.051	.076
				.076				.015	.03				.081
Flour:													
In barrels	per barrel	.304	.365	.487	.487	.487	.243	.243	.243	.243	.304	.304	.426
In sacks	per 2,240 lbs.	2.43					1.825	1.217	1.46	1.825	1.825		
Clover seed	do	2.73	3.04	3.04	3.04	2.73	1.825	1.217		1.825	1.825	2.13	3.04
Bacon	do	3.65	3.65	4.26	4.26	4.26	3.65	3.65	3.65	3.65	3.65	3.65	4.26
Hams	do	3.65	4.86	6.08	4.86	4.86	4.26	3.04	3.04	3.65	3.65	4.26	4.26
Lard	do	3.65	4.86	5.47	4.86	4.86	4.26	3.04	3.04	3.65	3.65	4.26	4.26
Lard, in small packages	do	4.86	4.86	6.69	4.86	4.86	4.26	3.04	3.04	3.04	3.04	4.26	4.26
Tallow	do	4.86	6.08	4.86	6.08	6.08	5.47	4.26	4.26	4.26	4.26	5.47	4.26
Beef	per tierce	.73	.973	1.217	.973	.973	.852	.008	.608	.73	.73	.73	.852

Commodity	Rate basis												
Pork	per barrel	.609	.487	.426	.425	.426	.426	.608	.669	.73	.852	.73	.487
Butter	per 2,240 lbs	7.30	6.08	6.08	6.08	6.08	6.08	6.08	7.30	7.30	8.51	8.51	6.08
Cheese	do	6.08	6.08	6.08	6.08	4.86	4.86	6.08	7.30	7.30	7.30	7.30	6.08
Cotton-seed meal	do	4.26	3.04	3.04	2.73	3.04	2.43	3.04	3.65	4.26	3.65	4.26	3.05
Cotton-seed oil	per barrel	.852	.73	.608	.608	.608	.608	.852	.973	.973	.973	.73	.73
Cotton	per 100 lbs	.973											
Hops	per 100 lbs	.75	.75	.75	.75	.75	.75	.75	.75	.275	.275	.75	.75
Lumber:													
Hard	per 2,240 lbs	3.65	4.26	3.65	3.65	3.65	3.65	4.26	4.26	4.86	4.86	4.86	3.65
Soft	do	4.26	4.26	4.86	4.26	4.26	4.26	4.86	4.86	5.47	4.86	5.47	4.26
Oil cake	per 280 lbs	1.825	.304	.243	1.825	1.46	1.217	1.70	2.73	3.04	3.04	3.04	2.43
Rosin	per 100 lbs	.243	.30	.20	.243	.182	.243	.304	.426	.04	.04	.487	.304
Tobacco	per 100 lbs	.20	.25	.25	.25	.25	.25						
Tobacco, in cases	per 40 cubic feet	4.86	4.66	4.86	4.86	4.86	4.86	4.86	4.86	4.86	4.86	4.86	4.86
Tobacco, in hogsheads	per hogshead	4.26	4.56	4.56	3.95	4.56	4.26	3.56	4.26	4.26	4.26	4.26	4.26
Apples:													
Green	per barrel	.73	.73	.71	3.04	.73	.73	.73	.73	.73	.73	.73	.73
Dried	per 40 cubic feet	3.65	3.04	3.04	3.04	3.04	3.34	3.65	3.65	4.26	3.65	3.65	3.65
Hay, pressed, in bales	per 2,240 lbs	6.08	8.51	8.51	4.86	4.56	4.56	3.56	4.86				
Cattle	per head				10.95	9.73							
Sheep	do				1.34	1.217							
Measurement	per ton of 40 cubic feet	4.86	4.86	4.26	3.04	3.65	3.65	3.65	3.65	3.65	4.26	3.65	3.65
Primage	per cent	5	5	5	5	5	5	5	5	5	5	5	5

NEW YORK TO NEWCASTLE VIA WILSON LINE.

Commodity	Rate basis												
Wheat	per 60 lbs	$0.091	$0.081	$0.086	$0.051	$0.051	$0.061	$0.061	$0.061	$0.061	$0.071	$0.071	$0.071
Corn	do	.091	.081	.086	.051		.061	.061	.061	.061	.071	.071	.071
Rye	do	.091	.081	.086	.051		.061	.061	.126	.071	.071	.071	.071
Oats	per 320 lbs	.487	.608	.588	.051		.456	.426	.061	.456	.487	.487	.456
Barley	per 48 lbs	.081	.081	.041			.061	.061	.061	.061	.071	.071	.071
Flour:													
In barrels	per barrel	.608	.487	.487	.365	.365	.365	.365	.365	.365	.487	.487	.365
In sacks	per ton	3.65	3.65	3.65	3.01	3.04	3.04	3.04	3.13	3.04	3.04	3.04	3.04
Corn meal, in sacks	do	3.65	3.65	4.26	3.04	3.04	3.04	3.04	2.43	3.04	3.04	b.365	3.04
Oatmeal, in sacks	do	3.04	3.26	4.26	3.01	3.04	3.04	3.04	2.43	3.65			4.20
Clover seed	do	4.26	4.66	4.86	4.86	4.86	3.65	3.65	3.65	3.65	3.65	3.65	4.26
Beef:													
Canned	do	3.65	4.86	4.86	3.65	3.65	3.65	3.65	3.65	3.65	3.65	3.65	4.26
In tierces	per tierce	3.73	4.97	4.871	3.73	3.73	3.73	3.73	3.71	3.73	3.67	3.71	3.872
Tallow	per ton	3.65	4.86	4.86	3.65	3.65	3.65	3.65	3.65	3.65	3.65	3.65	4.20
Pork	per barrel	.547	.73	.73	.487	.487	.487	.487	.487	.487	.547	.487	.608
Bacon	per ton	3.65	4.86	4.86	3.65	3.65	3.65	3.65	3.65	3.65	3.65	3.65	3.65
Hams	do	3.65	4.86	4.86	3.65	3.65	3.65	3.65	3.65	3.65	3.65	3.65	4.26

a Per ton. b Per barrel.

Transatlantic traffic rates—Continued.

NEW YORK TO NEWCASTLE VIA WILSON LINE—Continued.

Commodities.		1895.						1896.					
		Oct. 1.	Nov. 1.	Dec. 1.	Jan. 1.	Feb. 1.	Mar. 1.	Apr. 1.	May 1.	June 1.	July 1.	Aug. 1.	Sept. 1.
Lard:													
In tierces	per ton	$3.05	$4.86	$4.86	$4.86		$3.65	$3.65	$3.65	$3.65	$3.65	$3.65	$4.26
In small packages	do	4.26	5.47	5.47	5.47		4.26	4.26	4.26	4.26	4.26	4.26	4.86
Butter	do	7.30	7.30	7.30	6.08		7.30	5.47	5.47	6.08	6.08	6.08	6.08
Cheese	do	4.86	6.08	6.08	6.08		5.47	5.47	5.47	6.08	6.08	6.08	6.08
Cotton-seed meal	do	4.26	4.26	4.26	4.26								
Cotton-seed oil	per barrel	.73	.973	.973	.973		.73	.73	.73	.73	.73	.73	.852
Lumber:													
Hard	per ton	3.65	3.65	4.86	4.86		4.26	4.26	4.26	4.26	4.26	4.26	4.26
Soft	do	4.26	4.26	5.47	5.47		4.86	4.86	4.86	4.86	4.86	4.86	4.86
Oil cake	do	3.65	3.65	3.65	3.65		3.04	2.73	3.04	2.43	3.04	3.04	3.04
Rosin	per 280 lbs	.487	.547	.608	.608		.487	.487	.487	.487	.487	.487	.487
Tobacco:													
In cases	per 40 cubic feet	4.26					3.65	3.65	3.65				
In hogsheads	per hogshead			4.86			4.86	4.86	4.86	a.225	4.86	4.86	
Apples:													
Green	per barrel	.608	.608	.608	.608		.608	.608	.608	.608	.608	.608	.73
Dried	do	.608	.608	.608	.608		.608	.608	.608	.608	.608	.608	.73
Hay, pressed, in bales	per ton	4.86	4.86	8.51	6.08		6.08	6.08	6.08	6.08	6.08	6.08	
Cattle	per head	7.30	7.30		6.08		9.73	9.73	0.73	8.51	8.51	7.30	9.73
Sheep	do	.973	.973	.973	.973		.973	.973	.973	.973	.973		.973
Measurement	per ton of 40 cubic feet	3.65	3.65	4.86	4.86		3.65	3.65	3.65	{3.04 4.26}	3.65	3.65	3.65
Primage	per cent	5	5	5	5		5	5	5	5	5	5	

NEW YORK TO HULL VIA WILSON LINE.

Commodities.		1895.						1896.					
		Oct. 1.	Nov. 1.	Dec. 1.	Jan. 1.	Feb. 1.	Mar. 1.	Apr. 1.	May 1.	June 1.	July 1.	Aug. 1.	Sept. 1.
Wheat	per 60 lbs	$0.071	$0.081	$0.081	$0.081		$0.04	$0.046	$0.061	$0.061	$0.071	$0.071	$0.081
Corn	do	.071	.081	.081	.081		.04	.046	.061	.061	.071	.071	.081
Rye	do	.071	.081	.081	.081		.04	.046	.061	.061	.071	.071	.081
Oats	per 320 lbs	.487	.487	.608	.487				.426	.426	.487	.487	.487
Barley	per 48 lbs	.071	.081	.081	.081		.04	.046	.061	.061	.071	.071	.081
Flour:													
In barrels	per barrel	.487	.487	.487	.487		.365	.365	.243	.365	.426	.487	.487
In sacks	per ton	3.04	3.65	3.65	3.65		3.04	3.04	2.43	2.43	3.04	3.04	3.04
Corn meal, in sacks		3.04	3.65	3.65	4.26		3.04	3.04	2.43	2.43			
Oatmeal, in sacks	do	3.04	3.65	4.26	3.65		3.04	3.04	2.43	2.43	3.04		
Clover seed, in sacks	do	3.65	4.26	4.86	4.86		4.26	4.26	3.65	3.65			b.487

	Unit											
Beef:												
Canned	do	3.65	3.65	3.65	3.65	3.65	3.65	3.65	4.86	4.26	4.80	3.65
In tierces	per tierce	.73	.73	.73	.73	.487	.608	.973	.973	.487	.973	.73
Tallow	per ton	3.65	3.65	3.65	3.04	3.65	3.65	4.86	4.86	4.86	4.26	3.65
Pork	per barrel	.487	.487	.447	.426	.487	.487	.487	.609	.73	.73	.26
Bacon	per ton	3.65	3.04	3.04	2.43	2.43	2.43	3.65	4.89	4.86	4.86	3.487
Hams	do	3.65	3.04	3.04	2.43	2.43	2.43	3.65	4.86	4.86	4.86	3.65
Lard:												
In tierces	do	3.65	3.65	3.65	3.04	2.43	2.43	4.86	4.86	4.86	4.46	3.65
In small packages	do	4.26	3.65	3.04	3.04	2.43	2.41	4.86	5.47	5.47	5.47	3.65
Butter	do	6.08	6.08	3.04	6.08	4.86	4.86	6.08	6.08	7.30	7.30	7.30
Cheese	do	6.08	6.08	.487	4.86	.73	.487	.08	6.08	6.08	6.08	4.86
Cotton-seed meal	do				2.43	.73	.608	.73	4.26	.26	.26	4.26
Cotton-seed oil	per barrel	.73	.73	.73	.73	.73	.608	.73	.973	.973	.973	.73
Lumber:												
Hard	per ton	3.65	3.65	3.65	3.65	3.65	4.26	4.26	4.86	3.65	3.65	3.65
Soft	do	4.26	3.26	4.26	4.26	4.26	4.86	4.86	5.47	4.26	4.26	4.26
Oil cake	do	3.04	3.04	3.04	2.41	2.43	2.13	2.13	3.65	3.14	3.65	.547
Rosin	per 280 lbs	.608	.487	.487	.487	.365	.487	.487	.608	.608		
Tobacco:												
In cases	per 40 cubic feet					3.65	3.65	3.65	4.86	4.86	4.26	4.26
In hogsheads	per hogshead					4.86	4.26	4.26				
Apples:												
Green	per barrel	.73	.608	.608	.608	.608	.608	.608		.608	.608	.608
Dried	do	.008	.008	.008	.608	.608	.608	.608		.608	.608	.608
Hay, pressed, in bales	per ton	9.73	6.08	6.08	4.86	6.08	6.08	6.08	6.08	6.08	8.51	6.09
Cattle	per head	.73	9.73	.08	8.51	9.73	9.73	9.73	9.71	9.71	.973	.852
Sheep	do		.973		3.04	.973	.973	.973	.973	.973		
Measurement	per ton of 40 cubic feet	3.65	3.65	3.65	3.65	3.65	3.65	3.65	4.86	4.86	3.65	3.65
Primage	per cent.	5	5	5	5	5	5	5	5	5	5	5

NEW YORK TO HAVRE VIA COMPAGNIE GÉNÉRALE TRANSATLANTIQUE.

	Unit										
Flour:											
In barrels	per barrel	$0.80	$0.60	$0.60	$0.60	$0.60	$0.70	$0.70	$0.73	$0.70	$0.75
In sacks	per 100 lbs	.20	.20	.20	.20	.175	.18	.18	.18	.18	.20
Clover seed	do	.20	.225	.225	.225	.20	.225	.225	.18	.20	.20
Bacon	do	.30	.175	.225	.175	.30	.20	.20	.20	.20	.25
Hams	do	.25	.225	.225	.225	.30	.20	.20	.20	.20	.30
Lard	do	.30	.225	.275	.225	.30	.20	.20	.20	.20	.30
Lard, in small packages	do	.35	.275	.325	.325	.35	.25	.25	.25	.25	.40
Tallow	per tierce	.30	.225	.225	.20	.30	.20	.20	.20	.20	.50
Beef	per barrel	1.20	1.10	1.10	1.20	1.10	1.10	1.10	1.10	1.25	1.30
Pork	do	1.00	1.00	1.00	1.00	1.00	1.00	1.00	1.00	1.00	1.10
Butter	per 100 lbs	.40	.375	.375	.375	.40	.25	.25	.25	.40	.40

a Per 100 pounds.   *b* Per barrel.

Transatlantic traffic rates—Continued.

NEW YORK TO HAVRE VIA COMPAGNIE GÉNÉRALE TRANSATLANTIQUE—Continued.

Commodities	1895			1896								
	Oct. 1	Nov. 1	Dec. 1	Jan. 1	Feb. 1	Mar. 1	Apr. 1	May 1	June 1	July 1	Aug. 1	Sept. 1
Cheese..........per 100 lbs	$0.35	$0.20	$0.25	$0.375	$0.40	$0.25	$0.225	$0.25	$0.40	$0.40	$.40	$0.25
Cotton-seed meal........do	.25	.225	.25	.20	.25	.225	.225	.225	.25	.25	.25	.25
Cotton-seed oil..........do	.25	.25	.25	.25	.225	.25	.25	.25	.25	.20	.25	.30
Cotton...............do	.25	.25	.25	.25	.25	.25	.24	.24	.30	.225	.25	.30
Hops................do	.80	.90	.80	.90	.80	1.00	1.00	1.00	.80	1.30	.75	.90
Lumber:												
Hard..........per 2,240 lbs	6.00	6.00	7.00	7.00	6.00	8.00	7.00	7.00	8.00	a.30	7.00	8.00
Soft.............do	7.00	8.00	8.00	8.00	7.00	8.00	7.00	7.00	10.00	a.50	8.00	10.00
Oil cake........per 100 lbs	.15 / .175	.175	.20	.20	.175	.25	.225	.20	.20	.20	.175	.20
Tobacco:												
In cases......per 40 cubic feet	8.00	6.00	6.00	8.00	6.00	8.00	8.00	8.00	6.00	8.00	6.00	6.00
Maryland......per hogshead	4.00	3.50	4.00	4.00	3.50	4.00	4.00	4.00	4.00	3.56	4.00	4.00
Virginia.........do	5.00	5.00	5.00	5.00	5.00	4.50	4.50	4.50	5.00	4.00	5.00	5.00
Kentucky........do	6.00	6.00	6.00	6.00	6.00	5.00	5.00	5.00	6.00	5.00	6.00	6.00
Apples, dried.....per barrel	.60	.60	.60	.60	.60	.60	.60	.60	.75	.60	.60	.75
Measurement..per ton of 40 cubic feet	6.00	6.00	6.00	6.00	6.00	10.00	10.00	10.00	6.00	10.00	5.00	6.00
Primago..........per cent	8.00	10.00	10.00	10.00	8.00	10.00	10.00	10.00	10.00	10.00	10.00	10.00
	5	5	5	5	5	5	5	5	5	5	5	5

NEW YORK TO ANTWERP VIA WILSON LINE.

Commodities	1895			1896								
	Oct. 1	Nov. 1	Dec. 1	Jan. 1	Feb. 1	Mar. 1	Apr. 1	May 1	June 1	July 1	Aug. 1	Sept. 1
Wheat..........per 60 lbs	$0.051	$0.061	$0.066	$0.076		$1.03	$0.025	$0.051	$0.051	$0.061	$0.061	$0.086
Corn............do	.051	.061	.066	.076		.03	.025	.051	.651	.061	.061	.086
Rye.............do	.051	.061	.066	.076		.03	.025	.051	.051	.061	.061	.086
Oats..........per 320 lbs	.365	.426	.608	.426		.03		.426	.426	.426	.426	.608
Barley.........per 48 lbs	.046	.056	.066	.071			.025	.051	.051	.061	.061	.086
Flour:												
In barrels.....per barrel	.365	.487	.487	.365		.243	.243	.243	.365	.365	.365	.426
In sacks......per ton	2.43	3.04	3.65	3.04		1.70	2.13	2.43	2.43	2.43	2.43	3.05
Corn meal, in sacks......do	2.43	3.04	3.65	3.65		1.70	2.13	2.43	2.43			
Oatmeal, in sacks.......do	2.43	3.04	4.96	3.65		1.70	2.13	2.43	2.43	2.43		
Clover seed, in sacks......do	3.04	3.65	4.86	4.26		3.65	3.65	3.65	3.65	2.43		6.426
Beef:												
Canned...........do	3.04	3.04	4.86	4.26		3.65	3.04	3.04	3.04	3.04	3.04	4.86
In tierces......per tierce	.008	.008	.852	.852		.73	.008	.008	.008	.608	.608	.973

Tallow .... per ton	3.04	3.04	3.04	3.04	3.04	3.04	3.65	3.04	3.04	3.04	3.04	3.04	3.04	4.86
Pork .... per barrel	.426	.426	.426	.305	.426	.487	.608	.608	.608	.608	.426	.426	.426	.73
Bacon .... per ton	3.04	3.04	3.04	3.04	3.04	3.04	3.65	3.04	3.04	3.04	3.04	3.04	3.04	4.86
Hams .... do	3.04	3.04	3.04	3.04	3.04	3.04	3.65	3.04	3.04	3.04	3.04	3.04	3.04	
Lard:														
In tierces .... do	3.04	3.04	3.04	3.04	3.05	3.04	3.04	3.04	3.04	3.04	3.04	3.04	3.04	4.86
In small packages .... do	3.65	3.65	3.65	3.65	3.65	3.65	3.65	4.26	4.26	3.65	4.26	4.26	3.65	5.47
Butter .... do	6.08	6.08	6.08	6.08	6.08	6.08	6.08	6.08	6.08	6.08	6.08	6.08	6.08	6.08
Cheese .... do	4.86	3.65	4.86	4.86	4.86	7.30	7.30	6.08	6.08	6.08	6.08	6.08	6.08	6.08
Cotton-seed meal .... per barrel	3.65	.852	.608	.608	.608	.608	.608	.608	.608	.608	.608	.608		.973
Cotton-seed oil .... per barrel	.608	.00⅜	.00⅛	.00¾	.00⁷⁄₁₆	.00¾	.00⁵⁄₁₂	.00¾	.00⁷⁄₁₆	.00⁷⁄₁₆	.00⁷⁄₁₆	.00⁷⁄₁₆	.00⁷⁄₁₆	.00⁷⁄₁₆
Cotton .... per lb	.00³⁄₁₆													
Lumber:														
Hard .... per ton	3.65	3.65	3.65	3.65	3.04	3.65	3.65	3.65	3.65	3.65	3.65	3.65	3.65	4.26
Soft .... do	4.26	3.65	3.04	4.26	3.65	4.26	4.26	4.26	4.26	4.26	4.26	4.26	4.26	4.86
Oil cake .... do	2.43	3.04	3.04	3.04	2.13	1.70	1.70	2.43	2.43	2.43	2.43	2.43	2.43	3.34
Rosin .... per 280 lbs	.365	.487	.487	.487	.304	.365	.426	.365	.365	.365	.365	.365	.365	.487
Tobacco:														
In cases .... per 40 cubic feet	3.65	4.86	4.86	3.65	3.04	3.04	3.65	3.04	a.225	4.86	4.86	4.86	4.86	6.08
In hogsheads .... per hogshead	4.86	4.86	4.86	4.86	4.26	4.26	4.26	4.26						
Apples:														
Green .... per barrel	.487	.487	.608	.608	.608	.608	.608	.608	.608	.608	.608	.608	.447	.73
Dried .... do	.487	.487	.608	.608	8.51	9.73							.487	.73
Hay, pressed, in bales .... per ton														
Cattle .... per head	1.095	.973	.973	.973	.973	.973	.973	.973	.973	.973	.973	.973	.973	9.73
Sheep .... do					2.92	3.04	3.04	3.65	3.04	3.65				.973
Measurement ... per ton of 40 cubic feet	2.43	4.26	4.26	4.26					3.65				3.65	3.65
Primage .... per cent	3.04													
	5	5	5	5	5	5	5	5	5	5	5	5	5	5

NEW YORK TO BREMEN VIA NORTH GERMAN LLOYD STEAMSHIP COMPANY.

	$0.09	$0.085	$0.10	$0.11	$0.10	$0.07	$0.06	$0.08	$0.10	$0.11	$0.10	$0.12
	.09	.085	.10	.11	.10	.07	.06	.08	.10	.11	.10	.12
Wheat .... per 100 lbs	.09	.085	.10	.11	.10	.07	.06	.08	.10	.11	.10	.12
Corn .... do	.09	.085	.10	.11	.10	.07	.06	.08	.10	.11	.10	.12
Flour:												
In barrels .... per barrel	.50	.50	.50	.50	.50	.48	.48	.48	.65	.60	.60	.50
In sacks .... per 100 lbs	.15	.16	.16	.16	.16	.15	.14	.15	.15	.15	.15	.16
Clover seed .... do	.18	.20	.20	.22	.22	.18	.18	.18	.20	.18	.18	.22
Bacon .... do	.18	.20	.20	.22	.22	.18	.18	.18	.18	.18	.18	.22
Hams .... do	.18	.20	.20	.22	.22	.18	.18	.18	.18	.18	.18	.22
Lard .... do	.21	.23	.23	.25	.25	.21	.21	.21	.21	.21	.21	.25
Lard, in small packages .... do	.18	.20	.20	.22	.22	.18	.18	.18	.14	.18	.18	.23
Tallow .... do	1.00	1.10	1.10	1.10	1.00	.80	.80	.80	.18	a.18	a.18	a.20
Beef .... per tierce	.70	.75	.75	.78	.80	.72	.72	.72	.72	.72	.72	.72
Pork .... per barrel	.60	.60	.60	.60	.60	.60	.60	.60	.60	.60	.60	.60
Butter .... per 100 lbs												

a Per 100 pounds.   b Per barrel.

Transatlantic traffic rates—Continued.

NEW YORK TO BREMEN VIA NORTH GERMAN LLOYD STEAMSHIP COMPANY—Continued.

Commodities.		1895. Oct. 1	Nov. 1	Dec. 1	Jan. 1	1896. Feb. 1	Mar. 1	Apr. 1	May 1	June 1	July 1	Aug. 1	Sept. 1
Cheese	per 100 lbs	$0.30	$0.30	$0.30	$0.30	$0.36	$0.36	$0.36	$0.36	$0.36	$0.60	$0.60	$0.36
Cotton-seed meal	do	.16	.16	.18	.18	.16	.16	.16	.16	.16	.18	.18	.16
Cotton seed oil	per barrel	.90	.85	1.00	1.00	1.00	.80	.80	.80	a.80	.80	.80	.80
Cotton	per 100 lbs	.25	.25	.25	.25	.25	.25	.25	.18¾	.20	.20	.225	.25
Hops	do	.75	.75	.75	.75	.75	.75	.75	.75	.75	.75	.75	.75
Lumber:													
Hard	do	.20	.20	.20	.24	.22	.20	.20	.18	.20	.20	.20	.20
Soft	do	.24	.25	.25	.28	.28	.24	.22	.22	.24	.24	.25	.24
Oil cake	do	.16	.16	.16	.16	.16	.15	.15	.15	.15	.18	.18	.16
Rosin	per 280 lbs	.40	.40	.45	.50	.50	.40	.40	.40	.40	.50	.50	.50
Tobacco:													
In cases	per case	1.44	1.44	1.44	1.45	1.44	1.44	1.41	1.44	1.44	1.45	1.45	1.44
In hogsheads	per hogshead	6.00	6.00	6.00	5.00	4.20	4.20	4.20	4.20	5.40		6.00	4.20
Virginia	do	5.00	5.00	5.00	6.00	5.40	5.40	5.40	5.40	4.20	4.20	4.20	6.00
Kentucky	do	6.00	6.00	6.00							5.40	6.00	
Apples:													
Green	per barrel	.80	.80	1.00	1.00	.50	.65	.65	.65		1.00	1.00	.65
Dried	do	.50	.60	.70	.50	.40				.65	.65	.65	.60
Hay, pressed, in bales	per 100 lbs										.50	.50	
Measurement	per ton of 40 cubic foot	4.50	4.80	4.80	4.80	4.80	4.80	4.80	4.80	4.80	4.80	4.80	4.80
		9.80	9.60	9.60	9.60	9.60			6.00	6.00	}	8.00	10.00

NEW YORK TO COPENHAGEN VIA SCANDIA LINE.

		1895. Oct. 1	Nov. 1	Dec. 1	Jan. 1	1896. Feb. 1	Mar. 1	Apr. 1	May 1	June 1	July 1	Aug. 1	Sept. 1
Wheat	per quarter	$0.852	$0.882	$0.912	$0.912	$0.73	$0.304	$0.487	$0.487	$0.487	$0.669	$0.70	$0.912
Corn	do	.852	.882	.912	.912	.73	.365	.791	.578	.578	.669	.70	.912
Rye	do	.852	.882	.912	.912	.73	.304	.487	.487	.578	.669	.70	.912
Oats	per 320 lbs	b.791	b.821	b.852	b.852	b.660	.304	.487	.487	.487	.608	.70	.973
Barley	per quarter	.791	.821	.852	.852	.73	.365	.791	.578	.578	.608	.70	.973
Flour:													
In barrels	per barrel	.608	.608	.608	.608	.608	.608	.608	.547	.547	.547	.608	.73
In sacks	per ton	4.20	4.26	4.56	4.26	4.56	3.65	3.65	3.05	3.05	3.65	3.65	4.26

Corn meal ......do...	4.26	4.26	4.26	4.26	4.26	3.65	3.65	3.65	3.65	3.65	3.65	3.65	3.65	3.65	4.26
Oatmeal ......do...	4.26	4.26	4.26	4.26	4.26	3.65	3.65	3.65	3.65	3.65	3.65	3.65	3.65	3.65	4.26
Clover seed ......do...	6.08	4.56	4.56	6.08	6.08	5.47	5.47	4.26	4.26	4.26	4.56	4.56	4.56	4.86	5.47
Beef:															
Canned ......per 40 cubic feet.	6.08	6.08	6.08	6.08	6.08	5.47	5.47	4.56	4.26	4.26	4.56	4.56	4.86	4.86	5.47
Fresh ......per ton.	c.912	c.912	c.882	c.882	c.701	c.882	c.882	c.882	c.882						
In barrels ...	6.08	6.04	6.04	6.04	6.04	5.47	5.47	4.56	4.56	4.26	4.26	4.26	4.86	4.86	5.47
Tallow ......do...	6.08	6.08	6.08	6.08	6.08	5.47	5.47	4.56	4.26	4.26	4.56	4.56	4.86	4.86	5.47
Pork ......do...	6.08	6.08	6.08	6.08	6.08	5.47	5.47	4.56	4.26	4.26	4.56	4.56	4.86	4.86	5.47
Bacon ......do...	6.04	6.04	6.04	6.04	6.04	5.47	5.47	4.56	4.26	4.26	4.56	4.56	4.86	4.86	5.47
Hams ......do...	6.08	6.08	6.08	6.08	6.08	5.47	5.47	4.56	4.26	4.26	4.56	4.56	4.86	4.86	5.47
Lard:															
In tierces ......do...	6.08	6.08	6.08	6.08	6.04	5.47	5.47	4.26	4.26	4.26	4.56	4.56	4.86	4.86	5.47
In small packages ......do...	6.09	6.09	6.09	6.09	6.09	6.08	6.08	5.47	5.47	5.47	5.23	5.47	6.08	...	6.08
Butter ......do...	10.95	10.95	12.16	12.16	10.95	12.16	12.16	12.16	12.16	12.16	12.16	12.16	12.16	12.16	12.16
Cheese ......do...	10.95	10.95	12.16	12.16	10.95	12.16	12.16	12.16	12.16	12.16	12.16	3.65	12.16	12.16	12.16
Cotton seed meal, in sacks ......do...	1.56	4.56	4.86	4.56	4.56	3.65	3.65	3.65	3.65	3.65	.973	.973	.973	1.095	
Cotton-seed oil ......per barrel.	1.217	1.217	1.217	1.217	1.217	1.095	1.095	.972	.972	.973	.40		.973	.00	
Cotton ......per lb	.003½	.00⅞	.00¹⁄₁₀	.008		.00	.00	.00	.00	.40	1.00	1.00	.00		
Hops ......per 100 lbs.	1.00	1.00													
Hops ......per ton					9.73	7.30	7.30	7.30	7.30	7.30	7.30	7.30	7.30		7.30
Lumber:															
Hard ......do...	6.08	6.08	6.08	6.08	6.04	5.47	5.47	4.26	4.26	4.56	4.56	4.56	4.86	4.86	5.47
Soft ......per 40 cubic feet.	6.04	6.04	6.08	6.08	6.08	4.86	4.86	4.86	4.86	4.86	5.17	4.86	4.86	4.86	6.08
Oil cake ......per ton.	4.26	4.56	4.56	4.26	4.26	3.65	3.65	3.65	3.65	3.65	3.65	3.65	3.65	3.65	4.26
Rosin ......per 280 lbs.	.547	.608	.608	.669	.73	.73	.669	.669	.660	.487	.608	.008	.791		
Tobacco:															
In cases ......per 40 cubic feet.	6.08	6.08	6.08	6.08	6.08	4.86	4.86	6.08	6.08	6.08	6.04	6.04	6.04	6.04	6.08
In hogsheads ......per hogshead	6.69	7.30	7.30	4.86	6.86	6.86	6.69	6.83	4.86	4.86	4.86	5.47	5.47	4.86	4.86
Apples:															
Green ......per barrel.	1.00	1.00	1.00	1.00	1.00	.973	.973	.973	.973	.973	.973	.973			
Dried, in boxes ......per ton.	8.51	9.12	8.51	8.51	8.51	7.30	7.30	7.30	7.30	7.30	7.30	8.51			
Hay, pressed in bales ......do...	9.73	9.73	9.73			6.08	6.08	7.30	7.30	7.30	7.30	9.71			
Measurement ......per ton of 40 cubic feet.	6.08	6.04	6.04	6.08	6.08	6.08	6.08	6.08	6.08	6.08	6.08	6.08	6.08	6.08	6.08
Primage ......per cent.	5	5	5	5	5	5	5	5	5	5	5	5	5	5	5

a And 5 per cent primage   b Per quarter.   c Per barrel.

Transatlantic traffic rates—Continued.

NEW YORK TO COPENHAGEN VIA THINGVALLA LINE.

Commodities.		1895.						1896.					
		Oct. 1.	Nov. 1.	Dec. 1.	Jan. 1.	Feb. 1.	Mar. 1.	Apr. 1.	May 1.	June 1.	July 1.	Aug. 1.	Sept. 1.
Wheat	per quarter	$0.852	$0.882	$0.912	$0.912	$0.73	$0.365	$0.304	$0.547	$0.70	$0.608	$0.669	$0.912
Corn	do	.852	.882	.912	.912	.73	.365	.304	.547	.70	.608	.669	.912
Rye	do	.852	.882	.912	.912	.73	.365	.304	.547	.70	.608	.669	.912
Oats	do	.791	.821	.852	.852	a.70	a.304	a.274	a.481	a.608	a.547	a.547	a.791
Barley	do	.791	.821	.852	.852	.73	.365	.304	.608	.70	.608	.73	a.973
Flour:													
In barrels	per barrel	.608	.608	.608	.608	.608	.608	.608	.487	.487	.487	.487	.608
In sacks	per ton	4.25	4.26	4.56	4.56	4.56	3.65	3.65	3.65	1.52	3.65	3.95	4.26
Corn meal	do	4.25	4.26	4.56	4.56	4.56	3.65	3.65	3.65	1.52	3.65	3.95	4.20
Oatmeal	do	4.25	4.26	4.56	4.56	4.56	3.65	3.65	3.65	1.52	3.65	3.95	4.20
Clover seed	do	6.08	6.08	6.08	6.08	6.08	5.47	3.86	3.86	4.86	4.56	4.86	6.08
Beef:													
Canned	do	6.08	6.08	6.08	6.08	6.08	5.47	4.86	4.86	4.86	4.56	4.86	5.78
Fresh	per 40 cubic feet	6.08	6.08	6.08	6.08	6.08	6.08	6.08	6.08	6.08	6.08	6.08	6.08
In barrels	per barrel	.912	.912	.882	.882	.882	.791	.71	.73	.73	.73	.73	.852
Tallow	per ton	6.08	6.08	6.08	6.08	6.08	5.47	4.86	4.86	4.56	4.50	4.86	5.78
Pork	do	6.08	6.08	6.08	6.08	6.08	5.47	4.86	4.86	4.58	4.56	4.86	5.78
Bacon	do	6.08	6.08	6.08	6.08	6.08	5.47	4.86	4.86	4.56	4.56	4.86	5.78
Hams	do	6.08	6.08	6.08	6.08	6.08	5.47	4.86	4.86	4.56	4.56	4.86	5.78
Lard:													
In tierces	do	6.08	6.08	6.08	6.08	6.08	5.47	4.86	4.86	4.56	4.56	4.86	5.78
In small packages	do	6.69	6.69	6.69	6.69	6.69	6.08	5.47	5.47	5.17	5.17	5.47	6.39
Butter	do	10.95	10.95	12.16	12.16	12.16	12.16	12.16	12.16	12.16	12.16	12.16	12.16
Cheese	do	10.95	10.95	12.16	12.16	12.16	12.16	12.16	12.16	12.16	12.16	12.16	12.16
Cotton-seed meal	do	4.56	4.26	4.26	4.86	4.86	4.26	3.95	3.95	3.95	4.26	4.26	4.26
Cotton-seed oil	per barrel	1.217	1.217	1.217	1.217	1.217	1.095	.973	.973	.912	.912	.973	1.217
Cotton	per lb	.00¹⁄₈	.00¹⁄₈	.00¹⁄₈	.00¹⁄₈	.00¹⁄₈	.00¹⁄₈	.00¹⁄₈	.00¹⁄₈	.00¹⁄₈	.00¹⁄₈	.00¹⁄₈	.00⅜
Hops	per 100 lbs	.50	1.00										
Lumber:													
Hard	per ton	6.08	6.08	6.08	6.08	6.08	5.47	4.86	4.86	4.56	4.56	4.86	6.08
Soft	per 40 cubic feet	6.08	6.08	6.08	6.08	6.08	6.08	6.08	6.08	6.08	6.08	6.08	6.08
Oil cake	per ton	4.26	4.26	4.56	4.56	4.56	3.65	3.65	3.65	3.95	3.65	3.95	4.26
Rosin	per 280 lbs	.547	.578	.608	.547	.669	.487	.487	.497	.547	.487	.517	.508
Tobacco:													
In cases	per 40 cubic feet	6.08	6.08	6.08	6.08	6.08	6.08	6.08	6.08	6.04	6.08	6.04	6.08
In hogsheads	per hogshead	6.69	7.30	7.30	7.30	6.69	6.69	6.69	6.69	6.69	6.69	6.69	6.69
Apples:													
Green	per barrel	1.00	1.00	1.00	1.00	1.00	1.00	1.00	1.00	1.00	1.00	1.00	1.00
Dried, in boxes	per 40 cubic feet	6.08	6.08	9.12	6.08	6.08	6.08	6.08	6.08	6.08	6.08	6.08	8.51
Measurement	per ton of 40 cubic foot	6.08	6.08	6.08	6.08	6.08	6.08	6.08	6.08	6.08	6.08	6.08	6.08
Primage	per cent	5	5	5	5	5	5	5	5	5	5	5	5

BOSTON TO LIVERPOOL VIA CUNARD LINE.

		$0.04	$0.061	$0.061	$0.061	$0.061	$0.056	$0.036	$0.01	$0.036	$0.046	$0.036	$0.04	$0.071
Wheat	per bushel	.04	.961	.061	.061	.061	.056	.036	.012	.036	.046	.036	.04	.071
Corn	do	.04					.061	.036	.01				.04	.071
Rye	do		.071				.056		.012					.071
Oats	do						.061							.071
Barley	do	.051	.071	.071	.071	.071	.061	.036	.012 / .02	.04	.304	.04	.04	.071
Flour:														
In barrels	per barrel	.365	.365	.365	.365	.365	.365	.182	.122 / b .035	.243	.304	.243	.304	.243
In sacks	per ton	1.825	3.04	2.43	2.43	2.43	2.43	.973 / 1.217	b .01 / b .015	1.46	1.825	1.52	2.13	3.04
Corn meal	do	1.825	3.04	2.43	2.43	2.43	2.43	.973 / 1.217	b .04	1.46	1.825	1.52	2.13	3.04
Oatmeal	do	1.825	3.04	2.43	2.43	2.43	2.43	.973 / 1.217	b .035 / b .04	1.46	1.825	1.52	2.13	3.04
Clover seed	do	3.04	3.95	4.26	3.65	3.65	3.65	2.13	1.825	2.43	2.71	2.73	3.04	3.65
Beef:														
Canned	do	1.825 / 2.13	3.04	3.34 / 3.65	3.65	3.65	3.04	1.36	b .01	1.52	1.825	1.825	2.43	3.65
In barrels	per barrel	.304	.487	.487	.487	.487	.487	.243	.122 / b .03	.304	.304	.304	.365	.73
Tallow	per ton	1.825	3.04	3.34	3.04	3.04	3.04	1.36	b .04	1.52	1.825	1.825	2.43	3.65
Pork	do	1.825 / 2.13	3.04	3.34	3.65	3.65	3.04	1.36	b .01	1.52	1.825	1.825	2.43	3.65
Bacon	do	1.825	3.04	3.34	3.65	3.65	3.04	1.36	b .04	1.52	1.825	1.825	2.43	3.65
Hams	do	1.825	3.04	3.34	3.65	3.65	3.04	1.36	b .03	1.52	1.825	1.825	2.43	3.65
Lard:														
In tierces	do	1.825	3.04	3.16	3.04	3.04	2.73	1.217	b .03	1.46	1.825	1.52	2.13	3.34
In small packages	do	2.43	3.65	4.56	4.26	4.26	3.65	1.825	1.217	2.13	2.43	2.43	3.04	4.86
Butter	do	2.43	3.65	4.26	4.26	4.26	4.26	2.43	1.52 / 1.825	2.43	2.43	2.43	3.04	4.86
Cheese	do	2.43	3.65	4.26	4.26	3.65	3.65	3.65	1.52 / 1.825	2.13	2.43	2.43	3.04	4.86
Cotton-seed meal	do	2.13	3.34	3.04	2.71	3.34	2.43	1.217	.973	1.46	2.13	1.825	2.13	3.34
Cotton-seed oil	per barrel	.487	.73	.73	.73	.73	.73	.487	.243	.367	.487	.487	.608	.852
Cotton	per lbs	.00¾ / .00⅔	.20	.00⅔	.00⅔	.00⅔	b .21	b .13	b .07 / b .08	b .10	b .09½	b .00½	b .00½	.01⅔ / .01⅒
Hops	per 100 lbs		.40 / .46	.40	.40	.40	.40	.55	.15	.20	.25	.25	.30	.65
Lumber:														
Hard	per ton	2.13	3.04	3.34	3.04	3.04	3.04	1.825	1.58	1.825	2.43	2.13	2.43	3.65
Salt	do	2.40	3.65	3.65	3.65	3.65	3.65	3.04	2.43	3.04	3.65	3.65	3.65	4.86

a Per 320 pounds.   b Per 100 pounds.

Transatlantic traffic rates—Continued.

BOSTON TO LIVERPOOL VIA CUNARD LINE—Continued.

		1895.			1896.								
Commodities.		Oct. 1.	Nov. 1.	Dec. 1.	Jan. 1.	Feb. 1.	Mar. 1.	Apr. 1.	May 1.	June 1.	July 1.	Aug. 1.	Sept. 1.
Oil cake	per ton	$1.703	$2.73	$2.43	$2.19	$2.19	$0.973	{ a $0.03 / a.04 }	$1.46	$1.64	$1.46	$1.825	$2.73
Rosin	do	2.13				2.73	1.52	.97	1.52	1.825	1.825	2.13	2.13
Tobacco: In cases	per 40 cubic feet	5.47	2.43	2.43	2.43	2.43	1.825	1.217	1.825	2.43	1.825	1.825	
In hogsheads	per ton		6.08	6.69	6.08	6.08	4.26	3.04	3.65	4.86	4.86	4.26	2.43
Apples: Green	per barrel	.365	.365	.365	.365	.365	.365	.304	.365	.365	.365	{ .365 / .487 }	
Dried, in boxes	per 40 cubic feet	2.13	2.43	2.43	2.13	2.43	1.825	1.217	1.825	1.825	1.825	1.825	.487
Hay, pressed, in bales	per ton	3.65	6.08	6.69	6.08	6.08	3.65	2.43	3.65	3.65	3.04	1.825	2.43
Cattle	per head	7.30	7.30	7.30	9.73	11.56	10.95	9.73	9.73	9.73	9.73	3.65	8.51
Sheep	do	.852	.73	.73	1.217	1.217	{ .973 / 1.217 }	.973	1.095	1.217	.971	.73	9.73
Measurement	per ton of 40 cubic feet	{ 1.825 / 2.43 }	2.43	2.43	2.43	2.43	{ 1.217 / 1.825 }	{ 1.217 / 1.825 }	1.217	1.825	1.217	1.825	.73
Primage	per cent	5	5	5	5	5	5	5	5	5	5	5	2.43

BOSTON TO LIVERPOOL VIA WARREN LINE.

Commodities.		Oct. 1.	Nov. 1.	Dec. 1.	Jan. 1.	Feb. 1.	Mar. 1.	Apr. 1.	May 1.	June 1.	July 1.	Aug. 1.	Sept. 1.
Wheat	per bushel	$0.041	$0.061	$0.061	$0.061	$0.056	$0.015	$0.02	$0.04	$0.04	$0.04	$0.04	$0.071
Corn	do	.041	.061	.061	.061	.056	.015	.02	.04	.04	.04	.04	.071
Rye	do	.041	.061	.061	.061	.056	.015	.02	.04	.04	.04	.04	.071
Oats	do	.041	.061	.061	.061	.056	.015	.02	.04	.04	.04	.04	.071
Barley	do	.011	.061	.061	.061	.056	.015	.02	.04	.04	.01	.04	.071
Flour: In barrels	per 2,240 lbs	3.04	3.04	3.65	3.65	3.65	1.217	1.825	2.43	2.43	2.43	3.04	4.86
In sacks	do	1.825	3.04	2.73	2.43	2.43	1.217	.973	1.46	1.46	1.52	2.13	3.04
Corn meal, in sacks	do	1.825	3.04	2.73	2.43	2.43	1.217	.971	1:46	1.46	1.52	2.13	3.04
Oatmeal, in sacks	do	1.825	3.04	3.04	3.65	2.43	1.217	.973	1.46	1.46	1.52	2.13	3.04
Clover seed, in sacks	do	3.04	4.26	3.95	3.65	3.65	2.43	.973	1.46	1.46	1.52	2.13	3.04
Beef: Canned	do	3.05	2.43	3.65	3.65	3.04	1.46	1.94	2.43	2.43	2.43	3.34	4.26
Fresh	per 40 cubic feet		3.65	3.65	3.65	3.65	3.28	3.28	3.28	3.65	1.825	3.65	3.65
In barrels	per 2,240 lbs	3.05	2.43	3.04	3.04	3.04	1.46	.973	1.52	1.825	1.52	2.43	3.65
Tallow	do	1.825	2.43	3.04	3.04	2.73	1.34	.973	1.52	1.825	1.825	2.43	3.65
Pork	do	1.825	2.43	3.65	3.65	3.04	1.46	.973	1.52	1.825	1.825	2.43	3.65

Bacon	do	3.65	3.65	1.825	1.825	2.43	2.43	1.825	1.825	1.825	1.52	.973	1.46
Hams	do	3.65	3.65	1.825	1.825	2.43	2.43	1.825	1.825	1.825	1.52	.973	1.46
Lard:													
In tierces	do	3.04	3.04	1.825	1.825	2.13	2.43	1.825	1.825	1.825	1.52	.973	1.34
In small packages	do	4.86	4.86	2.43	2.43	3.04	3.04	2.43	2.43	2.13	2.13	1.46	1.947
Butter	do	4.26	4.26	2.43	2.43	3.04	3.04	2.43	2.43	2.13	2.13	1.825	1.947
Cheese	do	4.29	4.26	2.43	2.43	3.04	3.04	2.43	2.43	2.13	2.13	1.82	1.947
Cotton-seed meal	do	3.65	3.65	1.825	1.825	2.43	2.43	1.825	1.46	1.46	1.825	.973	1.46
Cotton-seed oil	do	3.65	3.65	2.43	2.43	2.43	1.46	1.825	1.825	1.825	1.825	.973	1.825
Cotton	per 100 lbs	.20	.20	.45	.197	.164	.262	.13	.17	.07	.15	.07	.12
Hops	do	.45	.45	.35	.35	.262	.164	.20	.17	.12	.15	.12	.25
Lumber:													
Hard	per 2,240 lbs	3.04	3.04	2.43	2.73	2.43	2.43	2.43	1.825	1.825	1.825	1.70	1.825
Soft	do	3.65	3.04	3.04	3.04	3.04	3.04	3.04	2.43	2.43	2.43	2.19	2.43
Oil cake	do	3.04	3.04	2.13	2.13	1.94	1.94	1.46	1.217	1.217	1.217	.973	.973
Rosin	do	3.04	3.04	2.43	2.73	2.43	2.13	1.825	1.46	1.46	1.52	1.217	1.217
Tobacco:													
In cases	do	4.86	4.86	4.86	4.86	4.86	3.65	3.65	2.43	2.43	3.04	2.43	2.41
In hogsheads	do	5.47	6.08	6.08	6.08	6.08	3.65	3.65	3.65	3.65	3.65	a .12	3.04
Apples:													
Green	per barrel	.365	.365	.365	.365	.365	.487	.365	.365	.365	.304	.304	.304
Dried	per 2,240 lbs	3.65	4.86	4.86	4.86	4.86	3.65	3.04	2.43	2.43	2.43	1.825	2.43
Hay, pressed, in bales	do	3.65	6.08	6.08	6.08	6.08	6.04	3.04	3.04	4.65	2.43	1.825	3.04
Cattle	per head	6.08	7.30	6.08	8.51	8.51	10.95	10.95	9.73	9.73	9.73	9.73	10.95
Sheep	do	.852	1.095	1.825	1.095	1.298	1.298	1.825	.973	.973	.973	1.095	1.095
Measurement	per ton of 40 cubic feet	1.825	1.825	1.825	1.825	1.825	.73	.73	1.825	1.46	1.217	1.217	1.46
Primage	per cent	5	5	5	5	5	5	5	5	5	5	5	5

BOSTON TO GLASGOW VIA ALLAN LINE ROYAL MAIL STEAMSHIPS

Wheat	per 60 lbs	$0.051	$0.061	$0.071	$0.071	$0.071	$0.036	$0.061	$0.051
Corn	do	.051	.061	.071	.071	.071	.036	.061	.051
Rye	do	.051	.061	.071	.071	.071	.036	.061	.051
Oats	per 32 lbs	.061	.081	.071	.061	.071	.02	.061	.051
Barley	per 48 lbs	.061	.096	.076	.071	.061	.016	.061	.036
Flour:									
In barrels	per barrel	.426	.608	.467	.487	.487	.365	.487	.487
In sacks	per top	2.73	3.04	3.65	3.65	2.73	1.217	2.73	1.947
Corn meal	do	2.73	3.04	3.65	3.65	2.73	1.217	2.73	1.947
Oatmeal	do	2.73	3.04	3.65	3.65	2.73	1.217	2.73	1.947
Clover seed	do	3.65	4.26	4.26	4.26	4.26	3.04	4.26	3.65
Beef:									
Canned	do	3.65	4.36	3.65	3.65	4.26	3.04	4.26	3.04
In barrels	per barrel	.73	.912	.973	.852	.912	.608	.912	.608
Tallow	per ton	3.65	4.26	4.26	3.65	4.26	3.04	4.26	3.04
Pork	do	3.65	4.56	4.26	4.86	4.36	3.04	4.36	3.04
Bacon	do	3.65	4.56	4.26	4.86	4.26	3.04	4.26	3.04
Hams	do	3.65	4.26	4.26	6.08	4.26	3.04	4.26	3.04

a Per 100 pounds.

Transatlantic traffic rates—Continued.

BOSTON TO GLASGOW VIA ALLAN LINE ROYAL MAIL STEAMSHIPS—Continued.

Commodities.	1895.			1896.								
	Oct. 1	Nov. 1	Dec. 1	Jan. 1	Feb. 1	Mar. 1	Apr. 1	May 1	June 1	July 1	Aug. 1	Sept. 1
Lard:												
In tierces .... per ton	$3.65	$4.26	$6.08	$4.86	$4.26	$3.04	$3.04	$3.04	$3.04	$3.04	$3.04	$4.26
In small packages .... do	4.26	5.47	6.69	6.09	4.86	3.65	3.65	3.65	3.65	3.65	3.65	5.47
Butter .... do	6.08	7.30	7.30	6.08	6.08	6.08	6.08	6.08	6.08	6.08	6.08	6.08
Cheese .... do	6.08	7.30	7.30	6.08	6.08	6.08	6.08	6.08	6.08	6.08	6.08	6.08
Cotton-seed meal .... do	3.04	3.04	3.65	3.65	3.04	1.46	1.46	1.46	1.70	2.19	2.19	3.04
Cotton-seed oil ...... per barrel	.487	.912	.852		.912				.73	.487	.608	.73
Cotton seed oil .... per ton						2.43	2.43	2.43				
Cotton .... per 100 lbs	.60	.75	.75	.75	.75	.75	.75	.75	.75	.75	.75	.75
Hops .... do	.75	.75	.75	.75	.75	.75	.75	.75	.75	.75	.75	.75
Lumber:												
Hard .... per ton	3.04	3.65	4.26	4.26	4.26	3.04	3.04	3.04	3.04	3.65	3.65	4.26
Soft .... do	3.65	4.26	4.86	4.86	4.86	3.65	3.65	3.65	3.65	4.26	4.26	4.44
Oil cake .... do	2.73	3.04	3.65	3.65	2.73	1.46	1.46	1.46	1.70	1.917	2.19	3.04
Rosin .... do	3.65	3.65	4.86	3.65	3.65	3.04	2.43	2.43	3.04	3.04	3.04	3.65
Tobacco:												
In cases .... do	6.08	6.08	6.08	7.30	4.86	6.08	6.08	6.08	6.08	6.08	6.08	6.08
In hogsheads .... do	6.08	6.08	6.08	7.30	4.86	6.08	6.08	6.08	6.08	6.08	6.08	6.08
Apples:												
Green .... per barrel	.73	.73	.73	.73	.73	.73	.73	.73	.73	.73	.73	.73
Dried .... do	.73	.73	.73	.73	.73	.73	.73	.73	.73	.73	.73	.73
Hay, pressed, in bales ...... per ton	6.08	6.08	6.08	7.30	6.08	6.08	6.08	6.08	6.08	6.08	6.08	6.08
Cattle .... per head	9.73	8.51	9.12	10.95	12.16	12.16	12.16	11.56	11.56	10.95	7.30	10.95
Sheep .... do	1.217	.973	1.217	1.217	1.46							
Measurement ...... per ton of 40 cubic feet.	3.65	3.65	3.65	3.65	3.65	3.77	3.04	3.04	3.65	3.65	3.65	3.65
Primage .... per cent.	5	5	5	5	5		5		5	5	5	5

BOSTON TO LONDON VIA FURNESS LINE.

Article	Unit												
Corn meal	do	2.73	2.73	2.31	2.13	1.825	1.46	1.217				2.73	3.35
Oatmeal	do	2.73	2.73	2.31	2.13	1.825	1.46	1.217				2.73	3.35
Clover seed	do	3.34	3.65	3.04	3.65				3.65	3.65	3.26	3.34	4.26
Beef:													
Canned	per 40 cubic feet												
Fresh	per 2,240 lbs	3.65	3.65	3.65	3.65	3.04	3.04	1.70	3.65	3.65	3.65	3.65	3.65
In barrels	per barrel			.365	3.65	3.28	3.28	3.28			a3.04		.852
Tallow	do	3.34	4.26	4.26	4.26			3.04	3.65	3.65	3.65	3.34	3.65
Pork	do	3.34	4.26	4.26	4.26	2.43	2.43	3.65	4.26	3.65	3.65	3.34	3.65
Bacon	do	3.34	4.26	4.26	4.26	2.43	2.43	3.65	4.26	3.65	3.65	3.34	3.65
Hams	do	3.34	4.26	4.26	4.26	2.43	2.43	3.65	4.26	3.65	3.65	3.34	3.65
Lard:													
In tierces	do	3.34	4.26	4.26	4.26	2.43	2.43	3.65	4.26	3.65	3.65	3.34	3.65
In small packages	do	3.65	4.86	8.51	4.86	3.65	3.65	4.26	4.86	4.26	4.26	3.65	4.26
Butter	do		4.26	4.04	4.26	4.26	4.80	3.65	4.86	4.86	4.86		4.86
Cheese	do	3.65	4.26	3.26	3.04	4.26	4.26	3.65	4.26	4.26	4.26		4.86
Cotton-seed meal	per barrel				.487			.608	.608	.487	.487		.852
Cotton-seed oil	per 100 lbs	.45	.53	.45	.45	.562	.50	.45	.55	.55	.487		.50
Hops	per 100 lbs												
Lumber:	2,240 lbs												
Hard	do	3.34	3.65	3.65	3.65	2.43	2.19	1.94	3.04	3.04	3.04	3.04	3.65
Soft	do	3.65	4.26	4.96	4.26	3.04	2.73	2.43	3.65	3.65	3.65	3.65	4.26
Oil cake	do	2.73	3.04	3.04	2.92	2.13	1.46	1.217	1.825	2.13	2.13	2.71	3.31
Rosin	do	2.43	3.04	3.04		3.04	1.46	1.34	1.825	2.13	2.13	2.74	3.65
Tobacco:													
In cases	per 40 cubic feet												
In hogsheads	per 2,240 lbs												
Apples:													
Green	per barrel	.487	.487	.487	.487	3.04	3.04	3.04	3.04	3.65	3.65	3.04	3.04
Dried	per 40 cubic feet					5.47	5.47	4.86	5.47	6.08	5.47	5.47	5.47
Hay, pressed, in bales	per 2,240 lbs	4.26	3.65	3.63	4.86	3.43	3.43	3.43	3.43	3.65	.608	.606	.487
Cattle	per head	6.08	6.08	7.30	9.73	9.73	9.73	9.73	9.73	8.51	8.51	3.04	3.04
Sheep	do	.852	.852	.973	1.095	1.217	1.217		1.217	0.95		1.217	9.73
Measurement	per ton of 40 cubic feet	3.04	3.65	3.04	3.04	2.43	2.43	2.43	1.217	1.217	1.217	1.217	8.51
									3.04	2.43		3.04	3.04
									3.01	3.04		3.65	3.65
Primage	per cent	5	5	5	5	5	5	5	5	5	5	5	5

BOSTON TO HAMBURG VIA HANSA-JOHNSTON LINE.

Article	Unit								
Wheat	per bushel	$0.119	$0.061	$0.091	$0.071	$0.101	$0.061	$0.107	$0.119	$0.143
Corn	do	.119	.061	.091	.071	.101	.061	.107	.119	.143
Rye	do	.119	.066		.071		.071		.119	.143
Oats	per quarter		.487		.73		.187		.73	
Barley	do	.487	.487		.669		.187		.73	

a Per 2,240 pounds.

Transatlantic traffic rates—Continued.

BOSTON TO HAMBURG VIA HANSA-JOHNSTON LINE—Continued.

Commodities		1895			1896								
		Oct. 1	Nov. 1	Dec. 1	Jan. 1	Feb. 1	Mar. 1	Apr. 1	May 1	June 1	July 1	Aug. 1	Sept. 1
Flour:													
In barrels	per ton	$3.65	$3.95	$4.26	$4.26	$1.52	$3.95	$3.95	$3.95	a $0.20	$3.65	$3.65	a $0.20
In sacks	do	3.34	3.65	3.65	3.65	3.65	3.40	3.40	3.40	a.15	3.40	3.34	a.15
Corn meal	do	3.65	3.65	3.65	4.26	3.65	3.65	3.65	3.65		3.65	3.65	a.17
Oatmeal	do	3.65	3.65	3.65	4.26	3.65	3.40	3.40	3.40		3.65	3.65	a.17
Clover seed	do	4.86	7.30	4.26	4.26	4.86	3.65	3.65	3.65	a.179	4.26	4.26	a.17
Beef:													
Canned	do	4.26	4.26	4.86	3.34	4.86	4.26	4.26	3.89	a.179	4.86	4.26	a.179
Fresh	per 40 cubic feet		3.04										
In tierces	per ton	4.26	4.26	4.86	4.86	4.86	4.26	4.26	4.56	a.179	4.86	4.26	a.179
Tallow	do	4.26	4.26	4.86	4.86	4.86	4.26	4.26	4.56	a.179	4.86	4.26	a.179
Pork	do	4.26	4.26	4.86	4.86	4.86	4.26	4.26	4.56	a.179	4.86	4.26	a.179
Bacon	do	4.26	4.26	4.86	4.86	4.86	4.26	4.26	4.56	a.179	4.86	4.26	a.179
Hams	do	4.26	4.26	4.86	4.86	4.86	4.26	4.26	4.56	a.179	4.86	4.26	a.179
Lard:													
In tierces	do	4.26	4.26	4.86	4.86	4.86	4.26	4.26	4.56	a.179	4.86	4.26	a.179
In small packages	do	4.56	4.88	6.08	5.47	5.47	4.86	4.86	6.09	a.215	5.47	4.86	a.215
Butter	do	10.34	4.66	9.73	7.30	9.73	9.73	9.73	9.73	a.477	a.477	9.73	a.477
Cheese	do	10.34	4.26	9.73	7.30	4.26	9.73	9.73	9.73	a.477	a.477	9.73	a.477
Cotton-seed meal	do	4.86	6.08		4.66		4.26	4.26	4.26		3.65	4.26	
Cotton-seed oil	do	.25	b 1.217	4.86	4.86	.25	.26	.26	.219	a.179	4.26	.26	a.179
Cotton	per 100 lbs	.25	.284	.50	.25	.25	.25	.25	.219	.23	.25	.25	.23
Hops	do												
Lumber:													
Hard	per ton	3.65	3.65	4.56	4.86	5.47	4.86	4.86	4.86	a.18	a.20	3.89	a.18
Soft	do	4.56	4.86	5.47	5.47	6.69	6.08	6.08	6.08	a.21	a.25	5.11	a.21
Oil cake	do	3.34	3.65	3.34	4.26	3.59	3.34	6.08	3.34	a.15	3.65	3.34	a.15
Rosin	do	3.65	3.95	3.65		3.65	b.973	b.973	b.973	3.339	3.339	3.409	3.339
Tobacco:													
In cases	do	7.30	6.08	6.08	5.10	2.30	7.30	7.30	7.30		a.596	6.08	6.69
In hogsheads	do	7.30	6.04	6.08	5.10	7.30	5.00	5.00	5.00	5.00	a.596	6.08	6.69
Apples:													
Green	per barrel	.73	.608	.973	.73	.73	.73	.73	.73	.75	.716	.73	.73
Dried	per ton	5.47	6.08	3.65	4.86	4.56				a.179	4.26	4.26	a.179
Hay, pressed, in bales	do	6.69	4.86	4.86									
Cattle	per head	8.51	2.30										
Sheep	do	1.46	1.217										
Measurement	per ton of 40 cubic feet	4.86	3.63	4.86	4.86	4.86	4.86	4.86	4.86	3.65	4.86	3.65	4.86
Primage	per cent	5	5	5	5	5	5	5	5	5	5	5	5

BALTIMORE TO LIVERPOOL VIA JOHNSTON LINE.

	$0.365	$0.365	$0.365	$0.304	$0.365	$0.487	$0.243	$0.243	$0.243	$0.274	$0.426	$0.365	$0.365	$0.497	$0.365
	.365	.365	.365	.304	.365	.487	.243	.243	.243	.274	.426	.365	.365	.497	.365
Wheat ....per quarter	.365														
Corn ....do															
Rye ....do															
Oats ....per 320 lbs	365														
Barley ....per 480 lbs	365														
Flour:															
In barrels ....per barrel	.10	.12	.135	.135	.14	.13	.30	.10	.10	.21	.243	.243	.243	.243	.14
In sacks ....per 100 lbs	.10	.12	.135	.135	.14	.13	.10	.10	.07	.07	.08	.10	.10	.115	.15
Corn meal ....do	.10	.12	.135	.135	.14	.13	.10	.10	.07	.07	.08	.10	.10	.115	.15
Oatmeal ....do	.10	.12	.135	.135	.14	.13	.10	.10	.07	.07	.08	.10	.10	.115	.15
Clover seed ....do	.15	.18	.19	.19	.19	.19	.16	.15	.15	.15	.15	.10	.15	.15	.17
Beef:															
Canned ....do	.08	.12	.10	.10	.16	.15	.10	.07	.07	.08	.10	.10	.10	.12	.14
In barrels ....do	.08	.12	.16	.16	.16	.15	.10	.07	.07	.08	.13	.13	.13	.12	.14
Tallow ....do	.08	.12	.16	.16	.16	.15	.10	.07	.07	.08	.12	.16	.16	.12	.14
Pork ....do	.08	.12	.16	.16	.16	.15	.10	.07	.07	.08	.12	.16	.16	.12	.14
Bacon ....do	.08	.12	.16	.16	.16	.15	.10	.07	.07	.08	.10	.10	.10	.12	.14
Hams ....do	.08	.12	.16	.16	.16	.15	.10	.07	.07	.08	.10	.10	.10	.12	.14
Lard:															
In tierces ....do	.08	.12	.16	.16	.16	.15	.10	.07	.07	.08	.10	.10	.10	.12	.17
In small packages ....do	.13	.16	.19	.19	.21	.18	.13	.10	.10	.13	.13	.13	.13	.15	.17
Butter ....do	.20	.20	.25	.25	.25	.23	.20	.12	.12	.12	.12	.16	.18	.18	.22
Cheese ....do	.20	.20	.25	.25	.25	.23	.20	.12	.12	.12	.12	.16	.19	.19	.22
Cotton-seed meal ....do															
Cotton-seed oil ....per barrel	.60	.63		.80			.60	.60	.60	.58	.10	.61	.61	.487	.30
Cotton ....per 100 lbs	.24	.22	.23	.28	.25		.20	.19	.19	.18	.15	.15	.15	.15	
Lumber:															
Hard ....do	.15	.14	.16	.17	.16	.16	.16	.15	.16	.13	.13	.13	.13	.13	.14
Soft ....do	.14	.15	.17	.10	.10	.17	.17	.16	.17	.14	.14	.14	.14	.14	.15
Oil cake ....do	.10	.12	.145	.14	.14	.13	.10	.07	.07	.08	.04	.10	.10	.115	.365
Rosin ....per 310 lbs	.304	.365	.365	.365	.365	.365	.243	.243	.243	.365	.243	.365	.365	.365	.365
Tobacco:															
In cases ....per 100 lbs	.20	.20	.25	.24	.24	.24	.23	.22	.22	.30	.29	.17	.17	.17	.23
In hogsheads ....do	.20	.20	.25	.25	.24	.24	.23	.22	.22	.20	.20	.17	.17	.17	.25
Apples:															
Green ....per barrel	.20	.70	.30	.65	.25	.65	.50	.50	.50	.50	.50	.50	.50	.50	.50
Dried ....do	.60	.30	.50	.30	.63	.70	.50	.50	.50	.50	.50	.4.86	.4.86	.3.65	.50
Hay, pressed, in bales ...per 2,240 lbs															
Cattle ....per head	7.30	6.08	7.30	7.30	9.12	10.95	9.73	9.73	9.73	3.65	3.65	4.86	4.86	3.65	9.61
Sheep ....do	.973	.973	1.217	1.217	1.217	1.46	1.095	1.095	1.095	9.73	9.73	9.73	8.51	2.43	
Measurement ....per ton of 40 cubic feet														2.43	
Primage ....per cent	5	5	5	5	5	5	5	5	5	5	5	5	5	5	5

a Per 100 pounds.   b Per barrel.

Transatlantic traffic rates—Continued.

BALTIMORE TO LONDON VIA ATLANTIC TRANSPORT LINE.

Commodities.		1895.			1896.								
		Oct. 1.	Nov. 1.	Dec. 1.	Jan. 1.	Feb. 1.	Mar. 1.	Apr. 1.	May 1.	June 1.	July 1.	Aug. 1.	Sept. 1.
Wheat	per 60 lbs	$0.061	$0.081	$0.081	$0.086	$0.081	$0.03	$0.04	$0.04	$0.061	$0.071	$0.065	$0.091
Corn	do	.061	.081	.081	.046	.081	.03	.04	.04	.061	.071	.045	.091
Rye	do	.061	.081	.081	.046	.081	.03	.04	.04	.061	.071	.045	.091
Oats	per 320 lbs	.426	.669	.669	.73	.660	.304	.406	.406	.70	.487	.487	.70
Barley	per 400 lbs	.487	.669	.669	.73	.660					.73	.487	.73
Flour:													
In barrels	per barrel	.426	.487	.487	.608	.155	a.19	a.19	a.19	.426	a.189	.426	.487
In sacks	per 100 lbs	.155	.169	.169	.169	.155	.112	.101	.101	b2.61	.126	b2.73	b2.53
Corn meal	do	.155	.169	.169	.169	.155	.112	.101	.101	b2.61	.126	b2.73	b2.53
Oatmeal	do	.155	.169	.169	.169	.155	.112	.101	.101	b2.61	.126	b2.73	b2.53
Clover seed	do	.183	.197	.197	.197	.183	.169	.169	.169	b3.04	.169	b3.04	b3.65
Beef:													
Canned	do	.183	.183		.169	.169	.14	.14	.14	b3.04	.14	b3.34	b3.83
In barrels	do	.211	.211	.183	.197	.197	.169	.169	.169	b2.73	.155	b3.34	b3.83
Tallow	do	.211	.211	.197	.197	.197	.169	.169	.169	b2.73	.155	b3.34	b3.83
Pork	do	.211	.211	.197	.197	.197	.169	.169	.169	b2.73	.155	b3.34	b3.83
Bacon	do	.211	.211	.197	.197	.197	.169	.169	.169	b2.73	.155	b3.34	b3.83
Hams	do	.211	.211	.197	.197	.197	.169	.169	.169	b2.73	.155	b3.34	b3.83
Lard:													
In tierces	do	.211	.211	.197	.197	.197	.169	.169	.169	b2.73	.155		b3.83
In small packages	do	.229	.236	.225	.225	.225	.197	.197	.197	b3.34	.181	b3.95	b4.44
Butter	do	.309	.35		.30		.281	.281	.281	.25	.25	.25	.25
Cheese	do	.281	.30	.169	.25	.225	.225	.225	.225	.25	.25	.25	.25
Cotton-seed meal	do	.158	.169		.169	.158	.112	.112	.112	b2.61	.126	.25	b3.53
Cotton-seed oil	per barrel	.73	a.197	.852	.973	.973	.547	.547	.547	.608	.608	b2.73	.852
Cotton	per 100 lbs	.25	.30	.25	.25	.25	.25	.25	.25	.25	.25	.25	.30
Hops	do	1.00	1.00	1.00	.75	.75	1.00	1.00	1.00	1.00	1.00	1.00	1.00
Lumber:													
Hard	do	.18	.197	.166	.197	.183	.169	.169	.169	b3.65	.169	b3.95	b4.26
Soft	do	.208	.225	.217	.225	.211	.197	.197	.197	b4.26	.197	b4.56	b4.86
Oil cake	do	.155	.169	.163	.169	.155	.112	.101	.101	b2.61	.126	b2.73	b3.53
Rosin	per 280 lbs	.365	.487		.487	.426	.365	.365	.365	.243	.487	.304	.426
Tobacco:													
In cases	per 100 lbs	.28	.35	.35	.30	.30	.22	.22	.22	.22	.21	.22	.35
In hogsheads	do	.25	.30	.30	.30	.25	.22	.22	.22	.21	.21	.21	.30
Apples:													
Green	per barrel	.73	.73	.197	c.197	.73	.487	.487	.487	.487	.73	.487	.73
Dried	per 100 lbs	.197	.19		.227	.197	.169	.169	.169	b3.14	.155	b3.34	b3.34
Hay, pressed, in bales	per ton	6.72	7.91	7.30	6.08	7.84	3.786	3.786	3.786	7.30	5.60	7.30	9.73
Cattle	per head	6.69	7.30	8.51	12.16	10.95	9.73	9.73	9.73	6.08	7.30	6.69	8.51
Sheep	do	.973	.973	1.005	1.46	1.46	1.217	1.217	1.217	.973	1.217	.973	1.217
Measurement	per ton of 40 cubic feet	3.04	3.65	3.65	4.26	3.65	3.04	3.04	3.04	3.04	3.04	3.04	3.65
Primage	per cent	5	5	5	5	5	5	5	5	5	5	6	5

BALTIMORE TO GLASGOW VIA BLUE CROSS-DONALDSON LINE.

		$0.046 .046	$0.08 .08	$0.061 .061	$0.08 .08	$0.053 .053 .053	$0.012 .042	$0.038 .038	$0.038 .038	$0.061 .061 .487 .061	$0.061 .061 .487 .061	$0.061 .061 .487 .061	$0.061 .061 .061 .061	$0.053 .053
Wheat	per bushel	.046	.08	.061	.08	.053	.30	.24	.24	c.304	c.243	c.365	c.365	.16
Corn	do	.046	.08	.061	.08	.053	.10	.08	.04	.10	.11	.125	.125	.16
Rye	do						.10	.08	.08	.10	.11	.125	.125	.16
Oats	per 320 lbs						.10	.08	.08	.10	.11	.125	.125	.18
Barley	per bushel						.18	.17	.15	.17	.17	.17	.17	
Flour:														
In barrels	per barrel	.15	.15	.15	.15	.15	.18	.17	.15	.15	.15	.15	.15	.18
In sacks	per 100 lbs	.15	.15	.15	.15	.15	.18	.17	.15	.15	.15	.15	.15	.18
Corn meal	do	.15	.15	.15	.15	.15	.18	.17	.15	.15	.15	.15	.15	.18
Oatmeal	do	.17	.18	.20	.20	.20	.18	.17	.15	.15	.15	.15	.15	.18
Clover seed	do						.18	.17	.15	.15	.15	.15	.15	.18
Beef:														
Canned	do	.15	.15	.19	.21	.26	.18	.17	.15	.15	.15	.15	.15	.23
In barrels	do	.20	.20	.24	.26	.26	.31	.22	.20	.20	.20	.20	.20	.23
Tallow	do	.28	.28	.30	.30	.28	.26	.25	.25	.25	.25	.25	.25	.25
Pork	do	.28	.28	.30	.30	.28	.26	.25	.25	.25	.25	.25	.25	.23
Bacon	do									.10	.11	.12		
Hams	do									a.15	c.608	c.608		
Lard:														
In tierces	do	.60	.65	.60	.65	.80	.70	.64	.60	.17	.15	.15	.15	.16
In small packages	do	.18	.18	.18	.19	.18	.17	.17	.17	.18	.17	.16	.16	.17
Butter	do	.19	.19	.19	.18	.19	.18	.18	.18	.10	.11	.125	.16	.16
Cheese	do	.15	.15	.15	.15	.15	.10	.08	.08	c.304	c.426	c.426	.365	.365
Cotton-seed meal	do	.365	.365		.487	.365	.365	.304	.365					
Cotton-seed oil	per barrel													
Lumber:														
Hard	per 100 lbs	.25	.25	.25	.25	.18	.17	.21	.23	.22	.20	.20	.21	.21
Soft	do	.25	.25	.25	.25	.19	.18	.21	.21	.21	.20	.20	.25	.25
Oil cake	do													
Rosin	per 310 lbs													
Tobacco:														
In cases	per 100 lbs	.30	.50	.25	.25	.70	.50	.50	.50	.50	.50	.50	.50	.50
In hogsheads	do	.60	.60	.65	.65	.70	.50	.50	.50	c4.86	r6.08	c4.86	c4.86	
Apples:														
Green	per barrel	8.51	7.30	9.73	9.73	12.16	9.73	9.73	9.73	12.16	10.95	9.73	9.73	10.34
Dried	do	1.46	1.46	1.46	1.46	1.46	1.46	1.46	1.46	1.46	1.46	1.46	1.46	
Hay, pressed, in bales	per ton									3.04	3.04	3.65	3.65	
Cattle	per head	5	5	5	5	5	5	5	5	5	5	5	5	5
Sheep	do													
Measurement	per ton of 40 cubic feet													
Primage	per cent.													

a Per 100 pounds.   b Per ton.   c And 5 per cent primage.

Transatlantic traffic rates—Continued.

BALTIMORE TO HAVRE VIA BLUE CROSS LINE.

Commodities.		1895.			1896.								
		Oct. 1.	Nov. 1.	Dec. 1.	Jan. 1.	Feb. 1.	Mar. 1.	Apr. 1.	May 1.	June 1.	July 1.	Aug. 1.	Sept. 1.
Wheat	per 60 lbs	$0.053	$0.061	$0 08	$0.076	$0.053	$0.03	$0.03	$0.03	$0.076	$0.068	$0.068	$0.056
Corn	do	.053	.061	.08	.076	.053	.03	.03	.03	.076	.068	.068	.056
Rye	do					.053	.03	.03	.03		.068	.068	
Oats	per 320 lbs			.487						.487			
Barley	per 60 lbs										.547	.547	.547
Flour:	per barrel												
In barrels	per barrel												
In sacks	per 100 lbs	.15	.15	.16	.16	.16	.45	.45	.45	.11	.14	.14	.16
Corn meal	do	.15	.15	.16	.16	.16	.15	.15	.15	.11	.11	.11	.16
Oatmeal	do	.15	.15	.16	.16	.16	.15	.15	.15	.11	.11	.11	.16
Clover seed	do	.20	.20	.21	.21	.21	.30	.30	.20	.20	.20	.20	.21
Beef:													
Canned	do	.18	.18	.20	.20	.20	.20	.20	.20	.18	.18	.18	.18
In barrels	do	.18	.18	.20	.20	.20	.30	.18	.18	.19	.18	.18	.18
Tallow	do	.18	.18	.20	.20	.20	.20	.18	.18	.18	.18	.18	.18
Pork	do	.18	.18	.20	.20	.20	.20	.18	.18	.18	.18	.18	.18
Bacon	do	.18	.18	.20	.20	.20	.20	.18	.18	.18	.18	.18	.18
Hams	do	.18	.18	.20	.20	.20	.20	.18	.18	.18	.18	.18	.18
Lard:													
In tierces	do	.18	.23	.20	.20	.25	.90	.18	.18	.18	.18	.18	.25
In small packages	do	.23	.23	.25	.25	.25	.25	.23	.23	.23	.23	.23	.25
Butter	do	.30	.30	.30	.30	.30	.30	.30	.30	.30	.30	.30	.30
Cheese	do	.30	.30	.30	.30	.30	.30	.30	.30	.30	.30	.30	.30
Cotton-seed meal	do									.14	.14	.14	
Cotton-seed oil	per barrel								.25	a .18	.85	.791	.32
Cotton	per 100 lbs	.25	.60	.30	.30	.30	.25	.25	.25	.22	.22	.22	
Lumber:													
Hard	do	.22	.20	.20	.21	.20	.20	.20	.20	.20	.20	.20	.20
Soft	do	.24	.21	.21	.22	.22	.21	.21	.21	.21	.21	.21	.21
Oil cake	do	.15	.18	.16	.16	.16	.15	.15	.15	.14	.14	.14	.16
Rosin	per 310 lbs						.365	.365	.365				.487
Tobacco:													
In cases	per 100 lbs	.25	.30	.30	.30	.30	.30	.30	.30	.30	.30	.30	.30
In hogsheads	do	.23	.30	.30	.30	.30	.30	.30	.30	.30	.30	.30	.30
Apples:													
Green	per barrel	.30	.70	.70	.25	.70	.70	.60	.60	.60	.60	.60	.70
Dried	do	.70	.70	.70	.70	.70	.70	.60	.60	.60	.69	.60	
Hay, pressed, in bales	per ton									4.86	7.30	7.30	
Measurement	per ton of 40 cubic feet									2.65	3.65	4.86	
Trimage	per cent	5	5	5			5	5	5	5	5	5	5

BALTIMORE TO BREMEN VIA NORTH GERMAN LLOYD STEAMSHIP COMPANY.

	$0.06	$0.095	$0.119	$0.131	$0.119	$0.107	$0.107	$0.107	$0.084	$0.107	$0.095	$0.107	$0.113
Wheat .... per 100 lbs.	.06	.095	.119	.131	.119	.107	.107	.107	.084	.107	.095	.107	.113
Corn .... do....	.06	.095	.119	.131	.119	.107	.107	.107	.084	.107	.095	.107	.113
Rye .... do....	.06	.095	.119	.131	.119	.107	.107	.107	.084	.107	.095	.107	.113
Oats .... per 320 lbs.												.42	.496
Barley .... per 100 lbs.				.22									.155
Flour: In barrels .... per barrel.	.40	.45	.45	.45	.40	.40	.40	.40	.40	.40	.45	.40	.396
In sacks .... per 100 lbs.	.15	.15	.15	.15	.15	.12	.12	.12	.12	.12	.15	.13	.13
Corn meal .... do....	.15	.15	.15	.15	.15	.12	.12	.12	.12	.12	.15	.13	.13
Oatmeal .... do....	.15	.15	.15	.15	.15	.12	.12	.12	.12	.12	.15	.13	.13
Clover seed .... do....	.19	.19	.19	.19	.19	.19	.19	.19	.19	.19	.19	.19	.24
Beef: Canned .... do....	.18	.20	.20	.22	.22	.18	.18	.18	.18	.18	.18	.18	.20
In barrels .... do....	.18	.20	.20	.22	.22	.18	.18	.1x	.18	.1x	.18	.18	.20
Tallow .... do....	.1x	.20	.20	.22	.22	.18	.18	.18	.18	.18	.18	.18	.20
Pork .... do....	.18	.20	.20	.22	.22	.18	.18	.18	.18	.18	.18	.18	.20
Bacon .... do....	.18	.20	.20	.22	.22	.18	.18	.18	.18	.18	.18	.18	.20
Hams .... do....	.18	.20	.20	.22	.22	.16	.18	.18	.18	.18	.18	.1x	.20
Lard: In tierces .... do....	.18	.20	.20	.22	.22	.18	.18	.18	.18	.18	.18	.18	.20
In small packages .... do....	.21	.23	.30	.25	.25	.21	.21	.21	.21	.21	.21	.21	.23
Butter .... do....	.25	.30	.30	.30		.30	.30	.30	.30		.12		.15
Cheese .... do....	.25	.30	.30	.13		.12	.12	.12	.12		.954		.35
Cotton-seed meal .... do....	.14	.15	.14	1.00	.15	.954	.954	.954	.954	.12		.13	.14
Cotton-seed oil .... per barrel.		.96	.954	.275	.954	.20	.20	.20	.20	.954	.954	.954	.954
Cotton .... per 100 lbs.	.25	.25	.23	.275	.275	.20	.20	.20	.20	.20	.18	.18	.26
Lumber: Hard .... do....	.18	.20	.20	.20	.20	.18	.18	.18	.18	.18	.18	.18	.20
Soft .... do....	.21	.23	.23	.33	.23	.21	.21	.21	.21	.21	.21	.21	.33
Oil cake .... do....	.14	.15	.14	.15	.14	.12	.12	.12	.12	.12	.12	.12	.13
Rosin .... per 310 lbs.	.258	.358	.426	.50	.477	.268	.268	.268	.268	.298	.298	.298	.477
Tobacco: In cases .... per case.	1.43	1.43	1.43	1.43	1.43	1.19	1.19	1.19	1.19	1.19	1.43	1.19	1.43
Virginia .... per hogshead.	4.77	4.70	4.77	4.70	4.17	4.17	4.17	4.17	4.17	4.77	4.17	4.77	4.17
Kentucky .... do....	5.96	6.00	5.96	6.00	5.36	5.36	5.36	5.36	5.36	5.36	5.36	5.36	5.96
Apples: Green .... per barrel.			1.00	1.00									.835
Dried .... per 100 lbs.	.20	.20	.20	.225	.225	.225	.225	.716	.716	.20	.20	.20	.225
Measurement .... per ton of 40 cubic feet.	4.77	4.77	3.58	4.77	4.77	4.77	4.77	4.77	4.77	5.36	3.58	4.77	3.58

a Per 100 pounds.

Transatlantic traffic rates—Continued.

BALTIMORE TO ANTWERP VIA PURITAN LINE.

Commodities.	1895. Oct. 1	Nov. 1	Dec. 1	1896. Jan. 1	Feb. 1	Mar. 1	Apr. 1	May 1	June 1	July 1	Aug. 1	Sept. 1
Wheat .... per 60 lbs	$0.053	$0.076	$0.11	$0.084	$0.053	$0.046	$0.038	$0.038	$0.061	$0.061	$0.061	$0.053
Corn .... do	.053	.076	.11	.084	.053	.046	.038	.038	.061	.061	.061	.053
Rye .... do					.053	.046	.038	.038	.061	.061	.061	
Oats .... per 320 lbs												
Barley .... per 60 lbs												
Flour:												
In barrels ...... per barrel						.36	.30	.245	.25	.243	.365	
In sacks .... per 100 lbs	.12	.15	.15	.15	.15	.12	.10	.085	.09	.10	.12	.12
Corn meal .... do	.12	.15	.15	.15	.15	.12	.10	.085	.09	.10	.12	.15
Oatmeal .... do	.12	.15	.15	.15	.15	.12	.10	.085	.09	.10	.12	.15
Clover seed .... do	.18	.18	.20	.20	.20	.19	.17	.17	.17	.17	.17	.18
Beef:												
Canned .... do	.15	.15	.18	.19	.19	.17	.14	.14	.14	.15	.15	.18
Fresh .... do	.15	.15	.18	.19	.19	.17	.14	.14	.14	.15	.15	.18
In barrels .... do	.15	.15	.18	.19	.19	.17	.14	.14	.14	.15	.15	.18
Tallow .... do	.15	.15	.18	.19	.19	.17	.14	.14	.14	.15	.15	.18
Pork .... do	.15	.15	.18	.19	.19	.17	.14	.14	.14	.15	.15	.18
Bacon .... do	.15	.15	.18	.20	.19	.17	.14	.14	.14	.17	.17	.18
Hams .... do	.15	.15	.18	.19	.19	.17	.14	.14	.14	.18	.18	.18
Lard:												
In tierces .... do	.15	.20	.18	.19	.19	.17	.14	.14	.14	.15	.20	.23
In small packages ..... do	.20	.20	.23	.24	.24	.22	.19	.19	.19	.20	.25	.23
Butter .... per barrel	.25	.25	.30	.30	.30	.25	.25	.25	.25	.25	.25	.25
Cheese .... per 100 lbs	.25	.25	.30	.30	.30	.25	.25	.25	.25	.25	.25	.25
Cotton-seed meal .. per barrel		.65						.65	.65			
Cotton-seed oil ... per 100 lbs								.20	a.14	.608	.608	
Cotton .... do	.25	.25	.25	.25	.25	.25	.20	.20	.18	.18	.18	.25
Lumber:												
Hard .... do	.20	.18	.18	.18	.18	.18	.18	.18	.17	.17	.17	.17
Soft .... do	.21	.20	.19	.19	.19	.19	.19	.19	.18	.18	.18	.18
Oil cake .... do	.12	.15	.15	.15	.15	.12	.10	.10	.09	.10	.12	.15
Rosin .... per 310 lbs	.426	.365			.487	.365	.304	.304	.243		.426	.487
Tobacco:												
In cases ...... per 100 lbs	.25	.25	.25	.25	.25	.30	.23	.23	.23	.20	.25	.25
In hogsheads .... do	.25	.25	.25	.25	.25	.30	.23	.23	.23	.20	.25	.25
Apples:												
Green .... per barrel	.30	.50	.70	.70	.70	.60	.60	.60	.60	.60	.60	.60
Dried .... do	.70	.50	.70	.70	.70	.60	.60	.60	.60	.60		
Hay, pressed, in bales .... per ton									4.86	6.08	4.86	
Measurement .. per ton of 40 cubic feet									3.04	3.04	3.65	
Primage .... per cent	5	5	5	5	5	5	5	5	5	5	5	5

a Per 100 pounds.

COASTWISE TRAFFIC RATES.

The following tables show the traffic rates between different ports of the United States for the commodities named from October 1, 1895, to October 1, 1896, by months:

BETWEEN NEW YORK AND NEWPORT NEWS, VA., VIA OLD DOMINION STEAMSHIP COMPANY.

Commodities.		1895			1896								
		Oct. 1	Nov. 1	Dec. 1	Jan. 1	Feb. 1	Mar. 1	Apr. 1	May 1	June 1	July 1	Aug. 1	Sept. 1
Apples, released:													
South bound	per barrel	$0.22	$0.32	$0.32	$0.32	$0.27	$0.27	$0.25	$0.25	$0.25	$0.25	$0.25	$0.25
North bound	do	.25	.25	.25	.25	.25	.25	.25	.25	.25	.25	.25	.25
Bacon and hams:													
In barrels	per 100 lbs.	.16	.16	.16	.16	.13	.13	.13	.13	.13	.13	.13	.13
In cases	do	.14	.14	.14	.14	.12	.12	.12	.12	.12	.12	.12	.12
Beef and pork:													
In barrels	do	.16	.16	.16	.16	.13	.13	.13	.13	.13	.13	.13	.13
In cases	do	.14	.14	.14	.14	.12	.12	.12	.12	.12	.12	.12	.12
Butter, released	do	.25	.25	.25	.25	.21	.21	.21	.21	.21	.21	.21	.21
Cheese, released	do	.22	.22	.22	.22	.18	.18	.18	.18	.18	.18	.18	.18
Cotton, compressed	per bale	.84	.81	.84	.84	.60	.60	.60	.60	.60	.60	.60	.60
Fertilizers, C. L.	per 100 lbs.	.11	.11	.11	.11	.09	.09	.09	.09	.09	.09	.09	.09
Flour:													
C. L.	per barrel	.28	.29	.28	.30	.23	.25	.23	.23	.23	.23	.23	.23
L. C. L.	do	.30	.30	.30	.30	.15	.15	.28	.28	.28	.28	.28	.28
Lard	per 100 lbs.	.17	.17	.17	.17	.14	.14	.14	.14	.14	.14	.14	.14
Lumber	per 1,000 feet	6.00	6.00	6.00	6.00	4.25	4.25	4.25	4.25	4.25	4.25	4.25	6.00
Potatoes:													
South bound	per barrel	.32	.32	.32	.32	.27	.27	.25	.25	.25	.25	.25	.25
North bound	do	.25	.25	.25	.25	.25	.25	.25	.25	.25	.25	.25	.25
Salt:													
In packages, C. L.	per 100 lbs.	.11	.11	.11	.11	.09	.09	.09	.09	.09	.09	.09	.09
In sacks, L. C. L.	do	.11	.14	.14	.14	.11	.11	.11	.11	.11	.11	.11	.11
Tobacco, unmanufactured:													
In hogsheads or tierces	do	.18	.18	.18	.18	.15	.21	.21	.15	.15	.15	.15	.15
In cases	do	.25	.25	.25	.25	.15	.15	.15	.15	.15	.15	.15	.15
Tobacco leaf, in cases (north bound)	do	.14	.14	.18	.18	.15	.15	.12	.13	.15	.12	.15	.15
Wheat, corn, rye, and oats	do	.14	.14	.14	.14	.12	.11	.11	.12	.12	.12	.12	.12
Barley	do	.11	.11	.11	.14	.11	.11	.11	.11	.11	.11	.11	.11
Livestock, O. R., released:													
Cattle	per head	9.90	9.90	9.90	9.90	8.40	8.40	8.50	8.50	8.50	8.50	8.50	8.50
Horses	do	9.90	9.90	9.50	9.90	8.40	8.40	8.50	8.50	8.50	8.50	8.50	8.50
Sheep	do	1.60	1.00	1.00	1.00	1.00	1.00	1.00	1.00	1.00	1.00	1.00	1.00
Hogs (weighing not over 400 pounds)	do	1.00	1.00	1.00	1.00	1.00	1.00	1.00	1.00	1.00	1.00	1.00	1.00

Coastwise traffic rates—Continued.

BETWEEN NEW YORK AND NORFOLK, VA., VIA OLD DOMINION STEAMSHIP COMPANY.

Commodities.		1895.			1896.								
		Oct. 1.	Nov. 1.	Dec. 1.	Jan. 1.	Feb. 1.	Mar. 1.	Apr. 1.	May 1.	June 1.	July 1.	Aug. 1.	Sept. 1.
Apples, released:													
South bound	per barrel	$0.27	$0.27	$0.27	$0.27	$0.27	$0.27	$0.25	$0.25	$0.25	$0.25	$0.25	$0.25
North bound	do	.25	.25	.25	.25	.25	.25	.25	.25	.25	.25	.25	.25
Bacon and hams:													
In barrels	per 100 lbs	.13	.13	.13	.13	.13	.13	.13	.13	.13	.13	.13	.13
In cases	do	.12	.12	.12	.12	.12	.12	.12	.12	.12	.12	.12	.12
Beef and pork:													
In barrels	do	.13	.13	.13	.13	.13	.13	.13	.13	.13	.13	.13	.13
In cases	do	.12	.12	.12	.12	.12	.12	.12	.12	.12	.12	.12	.12
Butter, released	do	.21	.21	.21	.21	.21	.21	.21	.21	.21	.21	.21	.21
Cheese, released	do	.18	.18	.18	.18	.18	.18	.18	.18	.18	.18	.18	.18
Cotton, compressed	per bale	.60	.60	.60	.60	.60	.60	.60	.60	.60	.60	.60	.60
Fertilizers, C. L.	per 100 lbs	.09	.09	.09	.09	.09	.09	.09	.09	.09	.09	.09	.09
Flour:													
C. L.	per barrel	.23	.23	.23	.23	.23	.23	.23	.23	.23	.23	.23	.23
L. C. L	do	.25	.25	.25	.25	.25	.25	.28	.28	.28	.28	.28	.28
Lard	per 100 lbs	.14	.14	.14	.14	.14	.14	.14	.14	.14	.14	.14	.14
Lumber	per 1,000 feet	4.25	4.25	4.25	4.25	4.25	4.25	4.25	4.25	4.25	4.25	4.25	4.25
Potatoes:													
South bound	per barrel	.27	.27	.27	.27	.27	.27	.25	.25	.25	.25	.25	.25
North bound	do	.25	.25	.25	.25	.25	.25	.25	.25	.25	.25	.25	.25
Salt:													
In packages, C. L.	per 100 lbs	.09	.09	.09	.09	.09	.09	.09	.09	.09	.09	.09	.09
In sacks, L. C. L	do	.11	.11	.11	.11	.11	.11	.11	.11	.11	.11	.11	.11
Tobacco, unmanufactured:													
In hogsheads or tierces	do	.15	.15	.15	.15	.15	.15	.15	.15	.15	.15	.15	.15
In cases	do	.21	.21	.21	.21	.21	.21	.21	.21	.21	.21	.21	.21
Tobacco leaf, in cases (north bound)	do	.15	.15	.15	.15	.15	.15	.15	.15	.15	.15	.15	.15
Wheat, corn, rye, and oats	do	.12	.12	.12	.12	.12	.12	.12	.12	.12	.12	.12	.12
Barley	do	.11	.11	.11	.11	.11	.11	.11	.11	.11	.11	.11	.11
Live stock, O. R., released:													
Cattle	per head	8.40	8.40	8.40	8.40	8.40	8.40	8.50	8.50	8.50	8.50	8.50	8.50
Horses	do	8.40	8.40	8.40	8.40	8.40	8.40	8.50	8.50	8.50	8.50	8.50	8.50
Sheep	do	1.00	1.00	1.00	1.00	1.00	1.00	1.00	1.00	1.00	1.00	1.00	1.00
Hogs (weighing not over 400 pounds)	do	1.00	1.00	1.00	1.00	1.00	1.00	1.00	1.00	1.00	1.00	1.00	1.00

BETWEEN NEW YORK AND RICHMOND, VA., VIA OLD DOMINION STEAMSHIP COMPANY.

Apples, released:															
South bound ..... per barrel	$0.27	.27	.27	.27	.27	.27	.27	.27	.27	.25	$0.25	$0.25	$0.25	$0.25	$0.25
North bound ..... do.	.27	.27	.27	.27	.27	.27	.27	.27	.27	.25	.25	.25	.25	.25	.25
Bacon and hams:															
In barrels ..... per 100 lbs.	.13	.13	.13	.13	.13	.13	.13	.13	.13	.13	.13	.13	.13	.13	.13
In cases ..... do.	.13	.13	.13	.13	.13	.13	.13	.13	.12	.12	.12	.12	.12	.12	.12
Beef and pork:															
In barrels ..... do.	.13	.13	.13	.13	.13	.13	.13	.13	.13	.13	.13	.13	.13	.13	.13
In cases ..... do.	.13	.13	.13	.13	.13	.13	.13	.21	.12	.13	.13	.13	.13	.13	.13
Butter, released ..... do.	.24	.24	.24	.24	.24	.24	.24	.24	.21	.21	.21	.21	.21	.21	.21
Cheese, released ..... do.	.21	.21	.21	.21	.21	.21	.21	.18	.18	.18	.18	.18	.18	.18	.18
Cotton, compressed ..... per bale	.70	.70	.70	.70	.70	.70	.70	.70	.60	.60	.60	.60	.60	.60	.60
Fertilizers, C. L ..... per 100 lbs.	.10	.10	.10	.10	.10	.10	.10	.10	.09	.09	.09	.09	.09	.09	.09
Flour:															
C. L ..... per barrel	.23	.23	.23	.23	.23	.23	.23	.23	.23	.23	.23	.23	.23	.23	.23
L. C. L ..... do.	.28	.28	.28	.28	.28	.26	.28	.28	.28	.24	.28	.28	.28	.28	.28
Lard ..... per 100 lbs.	.15	.15	.15	.15	.15	.15	.15	.15	.14	.14	.14	.14	.14	.14	.14
Lumber ..... per 1,000 feet	6.00	6.00	6.00	6.00	6.00	6.00	6.00	6.00	6.00	6.00	6.00	6.00	6.00	4.25	4.25
Potatoes:															
South bound ..... per barrel	.27	.27	.27	.27	.27	.27	.27	.27	.27	.25	.25	.25	.25	.25	.25
North bound ..... do.	.27	.27	.27	.27	.27	.27	.27	.27	.27	.27	.25	.25	.25	.25	.25
Salt:															
In packages, C. L ..... per 100 lbs.	.10	.10	.10	.10	.10	.10	.10	.09	.09	.09	.09	.09	.09	.09	.09
In sacks, L. C. L ..... do.	.13	.13	.13	.13	.13	.13	.13	.11	.11	.11	.11	.11	.11	.11	.11
Tobacco, unmanufactured:															
In hogsheads or tierces ..... do.	.15	.15	.15	.15	.15	.15	.15	.15	.15	.15	.15	.15	.15	.15	.15
In cases ..... do.	.21	.21	.21	.21	.21	.21	.21	.21	.15	.15	.15	.15	.15	.15	.15
Tobacco leaf, in cases (north bound) ..... do.	.15	.15	.15	.15	.15	.15	.15	.12	.12	.15	.15	.15	.15	.12	.12
Wheat, corn, rye, and oats ..... do.	.12	.13	.12	.12	.12	.12	.12	.13	.11	.12	.12	.12	.12	.12	.12
Barley ..... do.	.13	.13	.13	.13	.13	.13	.13	.13	.13	.13	.13	.13	.13	.13	.13
Live stock, O. R., released:															
Cattle ..... per head	9.90	9.90	9.90	9.90	9.90	9.90	9.90	9.90	10.00	10.00	10.00	10.00	10.00	10.00	10.00
Horses ..... do.	9.90	9.90	9.90	9.90	9.90	9.90	9.90	9.90	10.00	10.00	10.00	10.00	10.00	10.00	10.00
Sheep ..... do.	1.00	1.00	1.00	1.00	1.00	1.00	1.00	1.00	1.00	1.00	1.00	1.00	1.00	1.00	1.00
Hogs (weighing not over 400 pounds) ..... do.	1.00	1.00	1.00	1.00	1.00	1.00	1.00	1.00	1.00	1.00	1.00	1.00	1.00	1.00	1.00

BETWEEN NEW YORK AND PETERSBURG, VA., VIA OLD DOMINION STEAMSHIP COMPANY.

Apples, released:										
South bound ..... per barrel	$0.27	.27	.27	.27	.27	.27	.27	.27	$0.27	$0.27
North bound ..... do.	.27	.27	.27	.27	.27	.27	.27	.27	.27	.27
Bacon and hams:										
In barrels ..... per 100 lbs.	.13	.13	.13	.13	.13	.13	.13	.13	.13	.13
In cases ..... do.	.13	.13	.13	.13	.13	.13	.13	.13	.13	.13

Coastwise traffic rates—Continued.

BETWEEN NEW YORK AND PETERSBURG, VA., VIA OLD DOMINION STEAMSHIP COMPANY—Continued.

Commodities.		1895 Oct. 1.	Nov. 1.	Dec. 1.	1896 Jan. 1.	Feb. 1.	Mar. 1.	Apr. 1.	May 1.	June 1.	July 1.	Aug. 1.	Sept. 1.
Beef and pork:													
In barrels	per 100 lbs.	$0.13	$0.13	$0.13	$0.13	$0.13	$0.13	$0.13	$0.13	$0.13	$0.13	$0.13	$0.13
In cases	do.	.13	.13	.13	.13	.13	.13	.13	.13	.13	.13	.13	.13
Butter, released	do.	.24	.24	.24	.24	.24	.24	.24	.24	.24	.24	.24	.24
Cheese, released	do.	.21	.21	.21	.21	.21	.21	.21	.21	.21	.21	.21	.21
Cotton, compressed	per bale	.75	.75	.75	.75	.75	.75	.75	.75	.75	.75	.75	.75
Fertilizers, C. L.	per 100 lbs.	.10	.10	.10	.10	.10	.10	.10	.10	.10	.10	.10	.10
Flour:													
C. L.	per barrel	.23	.23	.23	.23	.23	.23	.23	.23	.23	.23	.23	.23
L. C. L.	do.	.28	.28	.28	.28	.28	.28	.28	.28	.28	.28	.28	.28
Lard	per 100 lbs.	.15	.15	.15	.15	.15	.15	.15	.15	.15	.15	.15	.15
Lumber	per 1,000 feet.	6.00	6.00	6.00	6.00	6.00	6.00	6.00	6.00	6.00	6.00	6.00	6.00
Potatoes:													
South bound	per barrel.	.27	.27	.27	.27	.27	.27	.27	.27	.27	.27	.27	.27
North bound	do.	.27	.27	.27	.27	.27	.27	.27	.27	.27	.27	.27	.27
Salt:													
In packages, C. L.	per 100 lbs.	.10	.10	.10	.10	.10	.10	.10	.10	.10	.10	.10	.10
In sacks, L. C. L.	do.	.13	.13	.13	.13	.13	.13	.13	.13	.13	.13	.13	.13
Tobacco, unmanufactured:													
In hogsheads or tierces	do.	.15	.15	.15	.15	.15	.15	.15	.15	.15	.15	.15	.15
In cases	do.	.21	.21	.21	.21	.21	.21	.21	.21	.21	.21	.21	.21
Tobacco leaf, in cases (north bound)	do.	.21	.21	.21	.21	.21	.21	.21	.21	.21	.21	.21	.21
Wheat, corn, rye, and oats	do.	.12	.12	.12	.12	.12	.12	.12	.12	.12	.12	.12	.12
Barley	do.	.13	.13	.13	.13	.13	.13	.13	.13	.13	.13	.13	.13

FROM PROVIDENCE, R. I., TO PHILADELPHIA, PA., VIA WINSOR LINE.

Commodities.		1895 Oct. 1.	Nov. 1.	Dec. 1.	1896 Jan. 1.	Feb. 1.	Mar. 1.	Apr. 1.	May 1.	June 1.	July 1.	Aug. 1.	Sept. 1.
Apples	per barrel.	$0.25	$0.25	$0.25	$0.25	$0.25	$0.25	$0.25	$0.25	$0.25	$0.25	$0.25	$0.25
Bacon	per 100 lbs.	.15	.15	.15	.15	.15	.15	.16	.15	.15	.15	.15	.15
Beef:													
In barrels	per barrel.	.40	.40	.40	.40	.40	.40	.40	.40	.40	.40	.40	.40
In half barrels	per half barrel.	.25	.25	.25	.25	.25	.25	.25	.25	.25	.25	.25	.25
Butter	per 100 lbs.	.25	.25	.25	.25	.25	.25	.25	.25	.25	.25	.25	.25
Cheese	do.	.20	.20	.20	.20	.20	.20	.20	.20	.20	.20	.20	.20
Cotton	do.	.15	.15	.15	.15	.15	.15	.15	.15	.15	.15	.15	.15
Fertilizers	do.	.125	.125	.125	.125	.125	.125	.125	.125	.125	.125	.125	.125
Flour	per barrel.	.25	.25	.25	.25	.25	.25	.25	.25	.25	.25	.25	.25
Hams	per 100 lbs.	.15	.15	.15	.15	.15	.15	.15	.15	.15	.15	.15	.15
Hogs	do.	.15	.15	.15	.15	.15	.15	.15	.15	.15	.15	.15	.15
Hops	do.	.25	.25	.25	.25	.25	.25	.25	.25	.25	.25	.25	.25

Lard:															
In barrels or tierces	do	.15	.15	.15	.15	.15	.15	.15	.15	.15	.15	.15	.15	.15	.15
In small packages	do	.18	.18	.18	.18	.18	.18	.18	.18	.18	.18	.18	.18	.18	.18
Lumber:															
Hard	per 1,000 feet	7.00	7.00	7.00	7.00	7.00	7.00	7.00	7.00	7.00	7.00	7.00	7.00	7.00	7.00
Soft	do	5.00	5.00	5.00	5.00	5.00	5.00	5.00	5.00	5.00	5.00	5.00	5.00	5.00	5.00
Pork:															
In barrels	per barrel	.40	.40	.40	.40	.40	.40	.40	.40	.40	.40	.40	.40	.40	.40
In half barrels	per half barrel	.25	.25	.25	.25	.25	.25	.25	.25	.25	.25	.25	.25	.25	.25
Potatoes	per barrel	.25	.25	.25	.25	.25	.25	.25	.25	.25	.25	.25	.25	.25	.25
Salt, in sacks	per 100 lbs	.15	.15	.15	.15	.15	.15	.15	.15	.15	.15	.15	.15	.15	.15
Tobacco, unmanufactured, in hogsheads	do	.20	.20	.20	.20	.20	.20	.20	.20	.20	.20	.20	.20	.20	.20
Wool, in sacks	do	.20	.20	.20	.20	.20	.20	.20	.20	.20	.20	.20	.20	.20	.20
Wheat and rye	per bushel	.08	.08	.08	.08	.08	.08	.08	.08	.08	.08	.08	.08	.08	.08
Corn	do	.06	.06	.06	.06	.06	.06	.06	.06	.06	.06	.06	.06	.06	.06
Oats	do	.05	.05	.05	.05	.05	.05	.05	.05	.05	.05	.05	.05	.05	.05
Barley	per 100 lbs	.20	.20	.20	.20	.20	.20	.20	.20	.20	.20	.20	.20	.20	.20

BETWEEN BOSTON, MASS., AND PORTLAND, ME., VIA PORTLAND STEAM PACKET COMPANY.

Apples:															
Green, C.L	per barrel	$0.11	$0.11	$0.11	$0.11	$0.11	$0.11	$0.11	$0.11	$0.11	$0.11	$0.11	$0.11	$0.11	$0.11
Green, L.C.L	do	.15	.15	.15	.15	.15	.15	.15	.15	.15	.15	.15	.15	.15	.15
Dried	per 100 lbs	.10	.10	.10	.10	.10	.10	.10	.10	.10	.10	.10	.10	.10	.10
Beef	per barrel	.20	.20	.20	.20	.20	.20	.20	.20	.20	.20	.20	.20	.20	.20
Butter	per 100 lbs	.15	.15	.15	.15	.15	.15	.15	.15	.15	.15	.15	.15	.15	.15
Cheese:															
In boxes	do	.15	.15	.15	.15	.15	.15	.15	.15	.15	.15	.15	.15	.15	.15
In barrels	do	.10	.10	.10	.10	.10	.10	.10	.10	.10	.10	.10	.10	.10	.10
Cotton:															
C.L	do	.06	.06	.06	.06	.06	.06	.06	.06	.06	.06	.06	.06	.06	.06
L.C.L	do	.10	.10	.10	.10	.10	.10	.10	.10	.10	.10	.10	.10	.10	.10
Fertilizers:															
C.L	do	.06	.06	.06	.06	.06	.06	.06	.06	.06	.06	.06	.06	.06	.06
L.C.L	do	.10	.10	.10	.10	.10	.10	.10	.10	.10	.10	.10	.10	.10	.10
Flour:															
C.L	per barrel	.12	.12	.12	.12	.12	.12	.12	.12	.12	.12	.12	.12	.12	.12
L.C.L	do	.15	.15	.15	.15	.15	.15	.15	.15	.15	.15	.15	.15	.15	.15
Hams:															
Loose	per 100 lbs	.15	.15	.15	.15	.15	.15	.15	.15	.15	.15	.15	.15	.15	.15
In barrels or casks	do	.10	.10	.10	.10	.10	.10	.10	.10	.10	.10	.10	.10	.10	.10
Hogs, dressed	do	.10	.10	.10	.10	.10	.10	.10	.10	.10	.10	.10	.10	.10	.10
Lard	per tierce	.30	.30	.30	.30	.30	.30	.30	.30	.30	.30	.30	.30	.30	.30
Lumber:															
Hard	per 1,000 feet	3.00	3.00	3.00	3.00	3.00	3.00	3.00	3.00	3.00	3.00	3.00	3.00	3.00	3.00
Soft	do	2.00	2.00	2.00	2.00	2.00	2.00	2.00	2.00	2.00	2.00	2.00	2.00	2.00	2.00
Pork	per barrel	.20	.20	.20	.20	.20	.20	.20	.20	.20	.20	.20	.20	.20	.20

Coastwise traffic rates—Continued.

BETWEEN BOSTON, MASS., AND PORTLAND, ME, VIA PORTLAND STEAM PACKET COMPANY—Continued.

Commodities	1895 Oct. 1	Nov. 1	Dec. 1	1896 Jan. 1	Feb. 1	Mar. 1	Apr. 1	May 1	June 1	July 1	Aug. 1	Sept. 1
Potatoes:												
C. L. ....per barrel	$0.11	$0.11	$0.11	$0.11	$0.11	$0.11	$0.11	$0.11	$0.11	$0.11	$0.11	$0.11
L. C. L. ....do	.15	.15	.15	.15	.15	.15	.15	.15	.15	.15	.15	.15
Salt:												
in boxes ....per 100 pounds	.15	.15	.15	.15	.15	.15	.15	.15	.15	.15	.15	.15
in bags ....do	.10	.10	.10	.10	.10	.10	.10	.10	.10	.10	.10	.10
Tobacco, unmanufactured, in hogsheads ...do	.10	.10	.10	.10	.10	.10	.10	.10	.10	.10	.10	.10
Wool:												
Washed ....do	.30	.30	.30	.30	.30	.30	.30	.30	.30	.30	.30	.30
Compressed ....do	.10	.10	.10	.10	.10	.10	.10	.10	.10	.10	.10	.10
Not compressed ....do	.15	.15	.15	.15	.15	.15	.15	.15	.15	.15	.15	.15
Wheat, in barrels:												
C. L. ....per barrel	.12	.12	.12	.12	.12	.12	.12	.12	.12	.12	.12	.12
L. C. L. ....do	.15	.15	.15	.15	.15	.15	.15	.15	.15	.15	.15	.15
Wheat, corn, rye, oats, and barley, in bags, per 100 pounds	.10	.10	.10	.10	.10	.10	.10	.10	.10	.10	.10	.10
Barley ....per bushel	.056	.056	.056	.056	.056	.056	.056	.056	.056	.056	.056	.056
Live stock:												
Oxen ....per head	1.50	1.50	1.50	1.50	1.50	1.50	1.50	1.50	1.50	1.50	1.50	1.50
	2.50	2.50	2.50	2.50	2.50	2.50	2.50	2.50	2.50	2.50	2.50	2.50
Cows ....do	1.25	1.25	1.25	1.25	1.25	1.25	1.25	1.25	1.25	1.25	1.25	1.25
	1.50	1.50	1.50	1.50	1.50	1.50	1.50	1.50	1.50	1.50	1.50	1.50
Sheep ....do	.15	.15	.15	.15	.15	.15	.15	.15	.15	.15	.15	.15
	.50	.50	.50	.50	.50	.50	.50	.50	.50	.50	.50	.50

BETWEEN BALTIMORE, MD., AND BOSTON, MASS., AND PROVIDENCE, R. I., VIA MERCHANTS AND MINERS' TRANSPORTATION COMPANY

Commodities	Oct. 1	Nov. 1	Dec. 1	Jan. 1	Feb. 1	Mar. 1	Apr. 1	May 1	June 1	July 1	Aug. 1	Sept. 1
Apples ....per barrel	$0.25	$0.25	$0.25	$0.25	$0.25 / .35	$0.25	$0.25	$0.25	$0.25	$0.35	$0.35	$0.25
Bacon and hams .....per 100 lbs	.21	.18 / .21	.21	.21	.21	.21	.21	.21	.21	.21	.21	.18 / .21
Do ....per barrel	.60	.60	.60	.60	.60	.60	.60	.60	.60	.60	.60	.60
Do ....per box	.75	.75			.75					.75	.75	.75
Beef, in tierces ....per tierce	.60	.60	.60	.60					.60	.60	.60	.60
Do ....per 100 lbs	.21	.18 / .21			.21	.21	.21	.21	.21	.21	.21	.18 / .21
Butter ....do	.33	.38	.33	.33	.33	.33	.33	.33	.33	.33	.38	.33
Cheese ....do	.28	.28	.28	.28	.28	.21	.21	.21	.33	.33	.28	.28

Cotton, compressed	per bale	1.00	.75	.75	1.00	1.00	1.00	1.00	1.00	1.00	1.00	.75	.75	.75	.75
Fertilizers	per 100 lbs.	.15	2.50	2.50	.15	.15	.15	.15	.15	.15	.15	3.00	3.00	3.00	.15
Do.	per ton	.25	.25	.25	.25	.25	.25	.25	.25	.25	.25	.25	.25	.25	2.50
Flour	per barrel	.20	.20	.20	.25	.25	.25	.25	.25	.25	.25	.25	.25	.25	.25
Hogs, dressed	per 100 lbs.	.21	.21	.18	.18	.21	.21	.21	.21	.21	.21	.21	.21	.21	.18
Lard	do.														.21
Lumber:															
C. L.	do.	.18	.15	.15	.18	.15	.15	.15	.15	.15	.15	.15	.15	.15	.15
L. C. L	do.	a.60	.18	.18	a.60	.18	.18	.18	.18	.18	.18	.18	.18	.18	.18
Pork	do.	.35	a.60	.21	.35	.21	.21	.21	.21	.21	.21	.21	.21	.25	.25
Potatoes	per barrel	.15	.35	.25	.15	.25	.25	.25	.25	.25	.25	.25	.25	.25	.35
Salt	per 100 lbs.	d2.50	b.17	.18	.15	.21	.21	.21	.21	.21	.21	.21	.21	.18	b.17
Tobacco, unmanufactured	do.	.30	c.21	.21	d2.50	.21	.21	.21	.21	.21	.21	.21	.21	.38	c.21
Wool	do.	.12	.33	.33	.30	.38	.38	.38	.38	.38	.38	.38	.38	.38	.33
Wheat, corn, rye, and oats	do.		.38	c.07	.12	e.07	.18	.18	.18	.18	.18	.18	.18	.18	.38
Barley	do.	.12	c.07	.17	.12	.17	.18	.18	.18	.18	.18	.18	.18	.18	c.07
Live stock, O. R., released:															
Cattle	per head	15.00	9.00	9.00	15.00	9.00	9.00	9.00	9.00	9.00	9.00	9.00	9.00	9.00	9.00
Horses	do.	18.00	18.00	18.00	18.00	18.00	18.00	18.00	18.00	18.00	18.00	18.00	18.00	18.00	18.00
Sheep	do.	3.00	5.00	5.00	3.00	5.00	5.00	5.00	5.00	5.00	5.00	5.00	5.00	5.00	5.00
Hogs	do.	3.00	5.00		3.00										5.00

BETWEEN BALTIMORE, MD., AND NORFOLK, VA., VIA MERCHANTS AND MINERS' TRANSPORTATION COMPANY.

Apples	per barrel	$0.20	f$0.20	$0.10	$0.20	$0.20	$0.25	$0.25	$0.25	$0.10	$0.25	$0.25	$0.10	$0.25	f$0.11
Bacon and hams	per 100 lbs.	.14	.14	.14	.14	.10	.10	.10	.10	.11	.14	.10	.14	.10	f.06
Beef, in tierces	per 100 lbs.	.30	f.14	f.09	.30	f.09	f.10	f.10	f.10	f.09	.17	.19	.10	.14	f.06
Butter	per 100 lbs.	.20	.17	.24	.14	.24	.24	.24	.24	.24	.17	.20	.14	.14	.07
Cheese	do.	.14	.24	.14	.50	.45	.45	.45	.45	.45	.14	.14	.14	.14	.10
Cotton, compressed	per bale	.50	.11	.45	.08	.11	.09	.09	.09	.11	.45	.50	.45	.45	.15
Fertilizers	per 100 lbs.	.08	.18	.11	.15	.25	.15	.15	.15	.25	.11	.08	.11	h.07	b.03
Flour	per barrel	.15	.20	.25	.20	.17	.09	.09	.09	.17	.18	.15	.17	.15	.06
Hogs, dressed	per 100 lbs.	.20	.14	.17	.14	.14	.10	.10	.10	.10	.10	.20	.10	g2.20	b.07
Lard	do.	.14	.17	.14	.14						.07	.14	.14	.15	b.06
Lard	do.	.15	.10	.10	.10	.10	.10	.10	.10	.10	.10	.15	.10	.15	.03
Lumber, C. L.	do.	.14	.17	.09	.14	.09	.10	.10	.10	.09	.14	.14	.14	.09	.07
Pork	per barrel	.15	.09	.14	.15	.14	h4.00	h4.00	h4.00	.10	.09	.15	.09	.14	.04
Lumber, C. L.	do.	.30	.17	.09	.30	.10	.10	.10	.10	.14	.14	.30	.11	.11	f.06
Pork	per barrel	.20	f.14	f.10	.20	.22	f.10	f.10	f.10	.10	.25	.20	.25	.25	f.10
Potatoes	do.	.08	f.09	b.07	.08	.09	.09	.09	.09	.09	.09	.08	.09	.09	b.03
Salt	per 100 lbs.	.50	f.17	f.17	.50	.17	f.17	f.17	f.17	.17	.14	.50	.14	.14	f.05
Tobacco, unmanufactured	per hogshead														

a Per barrel.
b Carload.
c Less than carload.
d Per hogshead.
e Per bushel.
f Per 100 pounds.
g Per ton.
h Per 1,000 feet.

Coastwise traffic rates—Continued.

BETWEEN BALTIMORE, MD., AND NORFOLK, VA., VIA MERCHANTS AND MINERS' TRANSPORTATION COMPANY—Continued.

Commodities	1895 Oct. 1	1895 Nov. 1	1895 Dec. 1	1896 Jan. 1	1896 Feb. 1	1896 Mar. 1	1896 Apr. 1	1896 May 1	1896 June 1	1896 July 1	1896 Aug. 1	1896 Sept. 1
Wool ...per 100 lbs	$0.30	$0.14 / .19	$0.30	$0.30	.24	.24	.24	.24	$0.24	$0.24	$0.24	$0.03
Wheat, corn, rye, and oats ...do	.10	.07	.10	.10	.07	.12	.12	.12	.07	.07	.07	.03
Barley ...do	.10	.07			.07	.12	.12	.12	.07	.07	.07	.03
Live stock, O. R., released:												
Cattle ...per head	3.00		3.00	3.00	2.50	2.50	2.50	2.50	2.50	2.50		
Horses ...do	5.00	2.50	5.00	5.00	2.50	2.50	2.50	2.50	2.50	2.50	2.50	2.50
Sheep ...do	.50		.50	.50	2.00	2.50	2.50	2.50				
Hogs ...do												

BETWEEN BALTIMORE, MD., AND SAVANNAH, GA., VIA MERCHANTS AND MINERS TRANSPORTATION COMPANY.

Commodities	1895 Oct. 1	1895 Nov. 1	1895 Dec. 1	1896 Jan. 1	1896 Feb. 1	1896 Mar. 1	1896 Apr. 1	1896 May 1	1896 June 1	1896 July 1	1896 Aug. 1	1896 Sept. 1
Apples ...per barrel	$0.25	$0.25	$0.25	$0.25	.25	.25	.25	.25	$0.25	$0.25	$0.25	$0.25
Bacon and hams ...per 100 lbs	.15	.15	.15	.15	.15	.15	.15	.15	.15	.15	.15	.15
Beef, in tierces ...do	a.60	a.60	a.60	a.60	.50	.50	.50	.50	.15	.15	.15	.15
Butter ...do	.25	.30	.25	.25	.25	.50	.50	.50	.30	.25	.30	.30
Cheese ...do	1.00	.75	1.00	1.00	.75	.21	.21	.21	.25	.25	.25	.50
Cotton, compressed ...per ton	b.15	b.15	1.00	1.00	.75	.75	.75	.75	.75	.75	.75	.75
Fertilizers ...per ton	.25	2.00	b.15	b.15	b.10	2.00	2.00	2.00	2.00	2.00	2.00	2.00
Flour ...per barrel	.20	.25	b.25	b.15	.25	.25	.25	.25	.25	.25	.25	.25
Hogs, dressed ...per 100 lbs	.15	.20	.20	.20	.30							
Lard ...do	b.18	.15	.15	.15	.15	.15	.15	.15	.15	.15	.15	
Lumber ...per 1,000 feet	c.60	b.18	b.18	b.18	4.50	5.00	5.00	5.00	4.00	4.00	.15	5.00
Pork ...per barrel	.25	c.60	c.60	c.60	.25	.15	.15	.15	.15	.15	.15	.15
Potatoes ...do	.15	.25	.25	.25	.25	.15	.15	.25	.25	.25	.25	.25
Salt ...per 100 lbs	.15	d.125 / e.15	.15	.15	.15	.15	.15	.15	.15	.15	.18	e.15
Tobacco, unmanufactured ...do	f2.50	.25	f2.50	f2.50	.30	.30	.30	.30	.25	.25	.50	.25
Wool ...do	.30	.50	.30	.30	.50	.50	.50	.50	.50	.50	.50	.25
Wheat, corn, rye, and oats ...do	.12	.12	.12	.12	.12	.15	.15	.15	.12	.12	.12	.12
Barley ...do	.12	.12	.12	.12	.12	.15	.15	.15	.12	.12	.12	.12
Live stock, O. R., released:												
Cattle ...per head	15.00		15.00	15.00	12.00	15.00	15.00	15.00	18.00	9.00		
Horses ...do	18.00	18.00	18.00	18.00	18.00	18.00	18.00	18.00	18.00	18.00	18.00	18.00
Sheep ...do	3.00		3.00	3.00	7.00	8.00	8.00	8.00				
Hogs ...do	3.00		3.00	3.00	7.00	8.00	8.00	8.00				

a Per tierce.　　b Per 100 pounds.　　c Per barrel.　　d Carload.　　e Less than carload.　　f Per hogshead.

FREIGHT CHARGES IN ENGLAND ON AGRICULTURAL PRODUCTS.

THE COMPLAINTS OF THE ENGLISH FARMER.

Low prices for products of the farm and the orchard and a strenuous foreign competition have induced the English farmer to consider carefully in what direction he may effect economies. He has complained of the railroad charges, and his complaints have been the more vehement on account of the low rates given for imported products. The farmer in Devonshire has seen United States apples landed in Plymouth and carried thence past his door to London for a lower rate for the combined ocean and inland transit than he could obtain from his nearest station to London. The Hampshire farmer has seen the same thing in respect of goods landed at Southampton from the United States, from the Channel Islands, and from France. The farmers on the east coast have had the same experience. Bacon in vast quantities from Denmark, eggs, poultry, and butter from there and from France, fresh fruits from Germany on the north and from Italy on the south, fresh vegetables from Belgium, Normandy, and Brittany, and even fresh milk in considerable quantities have reached London and the great centers of the manufacturing cities in the north at a rate which the English farmer could not secure for a much shorter haul.

Attention has been drawn to this state of things more particularly of late, in consequence of the greatly increased imports of perishable fruits and vegetables, and of the minor products of the farm, such as fowls, eggs, etc., and a great deal of criticism has been leveled at railroads. They have been freely charged with encouraging foreign at the expense of home industries. Perhaps it was in part the result of this criticism that some of the railroads have made during the past year strenuous efforts to meet the agriculturists halfway. Some account of what they have done and tried to do may not be without value in the United States.

THE EFFORTS OF THE RAILROADS.

It ought first to be explained that express companies as we know them do not exist in England, and that the system of selling by farmers on the spot to traveling agents of commission houses is little practiced.

The general direction of the efforts of the companies was toward securing cooperation on the part of farmers in forwarding their products. The companies said that they could afford to carry imported articles cheaply because quantities were large and the traffic constant, and if the farmers of any district would combine and send their produce in a lot they should have every advantage as to rates and every facility as to dispatch. The London and Northwestern Railway Company, one of the most important systems of the country, has made considerable reductions in its charges on agricultural products during the past few years, and during the past year has made special efforts to encourage business in this direction. In July, 1895, the agents of the company in agricultural districts were instructed by circular to call upon all farmers and shippers or individuals likely to become shippers of farm produce, with the object of bringing to their personal attention the fact that special low rates would be quoted for all shipments of not less than 2 tons, whether these were from one shipper or procured by combination.

There was little response to this attempt to extend facilities, and renewed efforts were made. Canvassers were instructed to make thorough investigation for their districts as to the extent to which an outlet could be found at large centers for agricultural and dairy produce then disposed of locally. Low special rates and improved service to distant markets were offered for large quantities. Canvassers were instructed to report in detail the places and the people visited, and more than 70 were employed in this work alone. The result was practically *nil.* Inspectors were then sent out from headquarters, and over 1,000 farmers were interviewed. The result is officially summarized as follows: (*a*) Number in favor of combination exceedingly few and no general apparent desire to have the present system of dealing with their produce modified; (*b*) more than one-half of those seen showed absolute indifference in the matter, excepting that some took the opportunity to ask for lower rates subject to present conditions; (*c*) to a large extent the traffic is already provided for by low rates, as to which no complaint was made; (*d*) much of the produce, as might be expected, is taken by wagon road to neighboring market towns; (*e*) a considerable part is bought by "middlemen," who visit the districts and pay the railway charges; (*f*) generally there does not appear to be any really acute depression in the farming industry in the London and Northwestern districts, and most of the farmers do not seem to look upon reduced railway rates as a cure for any depression there may be.

THE LONDON AND NORTHWESTERN RAILWAY COMPANY.

The London and Northwestern Railway Company traverses the great industrial counties of the north, and its primary interest is not the carriage of agricultural produce. Nevertheless, it handles great quantities of dead meat between Liverpool, London, and Scotland, and taps several districts of large area which are almost entirely agricultural.

THE GREAT WESTERN RAILWAY COMPANY.

The Great Western Railway Company has 2,000 miles of track and is first in Great Britain in point of mileage. It traverses a country almost entirely agricultural, and its policy toward the agriculturist is of special interest. Its chairman, Lord Emlyn, is a man of great energy, and is regarded in the railroad world as possessing unusual administrative capacity. In April, 1896, he gathered a large number of agriculturist patrons of the line at a meeting for the purpose of discussing the question of traffic charges on the products of the farm.

He said to the representative people present on the occasion that the Great Western Railway had no wish whatever to give any undue advantage to the foreigner. Foreign competition, he continued, appeared to have gained its footing by adapting and applying itself to giving those who had to carry goods the traffic in large quantities and packed in a convenient form, so that they might be carried and handled cheaply. That applied to the great bulk of the foreign produce which competed with the productions of the British farmer. The problem was how far that could be met by the combined efforts of the railway companies and the British farmers.

The Great Western Railway received large consignments of goods from abroad. For instance, there was the meat trade from Birkenhead to London, which was carried at a through rate by the train load and was delivered at Smithfield market for a low rate of transit. The British agriculturist complained that, although he lived in the vicinity of the line, he could not get his meat carried at the same rate as that from foreign parts. The Great Western board wanted to meet that objection; but it had to be met in a practical way. In the first place, it was necessary that they should carry the meat at a rate which would pay the company, and then, in proportion to the quantity carried, as the consignment decreased in weight and quantity so the rate for carriage per unit of quantity increased. The Birkenhead to London rate was called a 3-ton rate, but it was in effect a train-load rate. A minimum train load was about 30 tons of 2,240 pounds each. It was taken to Smithfield market and hoisted there without any extra charge. That was the 25 shillings ($6.06) a ton rate. To the query of the Birkenhead man who asked what the company would carry his meat for he would say: "If you give us 3 tons of meat to carry we will give you as nearly as we can a 25-shilling rate." But the agriculturist said it was of no use talking about 3 tons, as he could not send such a large quantity to the railway company; if he sent 2 tons would a proportionate rate be charged, and if he could not send 2 tons would 1 be accepted? He (the chairman) thought that something of the kind could be done. The company did not want to offer a rate or fix a quantity with which the farmer could not deal. It was perhaps useless to ask the British agriculturist individually to send 30 tons or even 3 tons, but could not the farmers by combination aggregate a consignment of 3 or 2 tons, or even 1 ton? This was being done even now by some agriculturists, without there being any special recognized rate for such quantities. He would like to establish, if it were likely to bring about a good result, a definite rate table by which everyone would know what was the special rate. It would be a satisfactory thing to have a tabulated rate so that if a man handed in at any station 3 tons, 2 tons, or even 10 hundredweight, he would know what was the special rate for his shipment.

It was worthy of consideration whether the company could not try to get as near as possible to the foreign meat rate for quantities less than a train load. The directors could not give a 25-shilling ($6.06) rate, which was the train-load scale, but for 2 and 1 tons and 10 hundredweight a graduated scale might be adopted, which would be a feasible thing if it could be undertaken. It of course involved combination for collecting purposes. A lot and a load were very different things. A carload was what could be put into one car, but a 3-ton lot might be scattered in two or three cars. There was another point in connection with the 3 or 2 ton lots which he did not think was altogether known by agriculturists. Under the suggested arrangement it would not be necessary for the entire lot to be made up of meat, or vegetables, or butter, or poultry; but if in the aggregate the different classes of articles made up the 3 tons they would be entitled to have the 3-ton rate. That was an item of importance which should be generally known. In various districts this aggregation was being carried out at the present time. At Aylesbury, meat, poultry, eggs, butter, rabbits, and other things were collected, packed, and sent to market by a man who was paid a commission by those for whom he collected, and who received the usual agent's commission from the company. There was considerable difficulty in dealing with the carriage of fruit. The Great Western had to carry fruit very long distances and in very small quantities, and when carried in small quantities fruit was one of the least paying articles of traffic. Sometimes there was only 3 hundredweight or 4 hundredweight of fruit in a car, and what was wanted to make it pay was to increase the quantity carried per carload. The Great Western had half-parcel rates for perishable freight in existence now, and carried 28 pounds 50 miles for 7 pence (14 cents). As to the charge for straw, the difficulty was that only about 1 ton of straw could be carried in a car made to carry 10 tons of ordinary freight. If it could be compressed the difficulty would be lessened, but at present straw was one of the worst paying articles of traffic.

The question had been raised of running an excursion train for live stock in bulk. The company had always had two systems of charging, one per head and one per car; but the directors, he believed, had not gone into the question of having several cars in a lot, or carrying live stock in bulk. In fact, what the directors had been aiming at in the discussions which had taken place was to see whether in England the system could be carried out which had sprung up abroad in regard to the better handling of the traffic. England was flooded with foreign traffic because it was well packed and well handled, and it was necessary to meet that with combination of the same sort; for producers could not make railway companies carry at a loss. Freight would have to be well packed if the traders wanted the railway companies to carry it cheaply. As to giving preference to foreign freight, the board had no intention of doing so. Lord Jersey had suggested that there should be a market day for a pick-up van, and that there should be less unpacking. That system was adopted in regard to certain articles for Birkenhead, Bristol, and other places. As to demurrage, the company allowed two clear days for a car to stand at a station, but they had to take care to keep the sidings open for the convenience of traffic generally. In regard to the question whether any 3-ton consignment could be broken into small lots for different deliveries, he had to say that a small extra charge of 2 pence (4 cents) or 3 pence (6 cents) would be made for delivery. In reference to grain rates, the board had no desire that foreign grain should have a preferential rate as compared with that for English grain, and if any case were brought before them in which a preference had been given they would carefully consider it and make full inquiry. Wherever there were low rates for large consignments of foreign grain the board had endeavored to give a countervailing benefit to the English agriculturists by carrying oil cake, manure, and other articles at low rates. He believed the company had given great facilities to the traders who used the Great Western line, and they wanted now to do something which would be a practical benefit to the farmers. They did not want to advertise any "paper" scheme, but to do something which would be of practical use in alleviating the prevailing agricultural depression.

The comments of Lord Emlyn are summarized at some length, as they are a clear and practical statement of the views entertained by English railway directors generally upon the question of the carriage of agricultural products. The company has been working on the lines laid down by the chairman not only in the direction of special rates and facilities for larger consignments, but also in the direction of relieving the farmer as much as possible of all trouble in connection with his shipments. The company, in other words, is ready to undertake collection and distribution. Taking certain agricultural centers, they are willing, experimentally, to send out from their stations men who will go round with carts to the various farms in the district and receive the produce from the farmers without requiring the latter even to pack it. The collectors would have baskets with them for that purpose, and also cloths for wrapping up any poultry, pork, or similar perishable produce that might be handed to them. Thus collected, with absolutely no trouble to the farmer, the produce would be brought to London or some other principal consigning center, and if the grower had not already arranged with a dealer or a private purchaser, to whom the consignment might in that case be delivered direct, the company would take the consignment to Covent Garden, Smithfield, or to some large dealer, as might seem best, would dispose of it and collect the money, forwarding the proceeds, minus freight and other charges, to the farmer. On a limited scale such a system has been in vogue for twenty years in the Aylesbury and Thames districts, where the Great Western Railway Company deals with ducks in this way at a charge of about 2 pence (4 cents) each bird. But the proposal to extend the system to agricultural produce generally is quite a new development.

In reply to a letter addressed by me to the Great Western board
inquiring as to what encouragement it had had from farmers in its new
policy, I was informed "that the revision of this company's arrange-
ments in regard to the conveyance of agricultural produce has not yet
advanced to a stage which will enable me to furnish any information
that will be of practical value." While it is no doubt premature to
udge of the result of this company's policy, I have understood that
the company has reason to feel somewhat dissatisfied with the way in
which its advances have been met.

THE GREAT EASTERN RAILWAY COMPANY.

The Great Eastern Railway Company, serving a country rich in
agricultural output and particularly in garden produce and fruit, has
started a plan which works on different lines than those of the other
companies. Instead of encouraging combination among farmers with
the idea of shipping large quantities at lower rates, it has offered the
most complete facilities for bringing producer and consumer together.
The character of the farms on this line no doubt induced its new
departure. It simply suited its methods to the nature of its business.
Its scheme, if widely developed and taken up on other roads, may have
a most considerable effect on existing methods of distribution in London.
The plan consists in the printing and general distribution throughout
London of a list of farmers on its line who are prepared to consign
produce directly to consumers. The system was started on the first of
December, 1895, and has had such support up to the present time as to
justify the company in its adoption. Several hundred farmers are now
represented on the list which the company spreads broadcast over
London. The householder sends his order to such farmer as he may
choose for butter, eggs, poultry, vegetables, and farm produce generally. The farmer fills the order, packing the box himself and handing
it to the company for delivery. There is a uniform charge all over the
line of 8 cents for 20 pounds, irrespective of distance.

The limit of weight allowed is 60 pounds, and the company will take
a parcel of this weight from anywhere on its line to London for 24
cents and deliver the parcel in its own wagons without extra charge
within a radius of 3 miles of its central station. The consignments
are carried in express passenger trains and usually delivered on the
day of transmission. Not only does the consumer get fresh garden
produce at a minimum charge for carriage, but he saves the middle
man's London charges. The wholesale and the retail man are eliminated

The object and desires of the railroad company in inaugurating this
scheme may be gathered from the words of one of its leading officials.
He says:

Wo have a tremendous residental population along the London end of our line
and this scheme was largely devised in its interest, and in the interest of the farmer
as well, I may say. The farmers get better terms for their produce, while the reduc
tion in cost to the public is very considerable. When our scheme is more widel;

known it will, I believe, contribute largely to reviving farming in the eastern counties. Our list of farmers who have engaged to supply the London consumer ranges over Cambridgeshire, Essex, Hertfordshire, Huntingdonshire, Norfolk, and Suffolk, and we bring parcels of fish from Lowestoft and Yarmouth to London at much the same rates. At first there were difficulties about the packing, but we have surmounted that. To facilitate matters for the farmers we have had boxes manufactured which are on sale at all our stations at cost price. Our cheapest box, 10 inches in length, 7½ inches in breadth, and 3 inches in depth, is on sale at 1½ pence (3 cents), and our biggest box, a very capacious article indeed, costs 5 pence (10 cents).

The official, continuing, says:

The Londoner is already taking very kindly to the scheme. I have evidence that it has saved money to many an anxious, overburdened London householder, who by its means gets his larder stocked with necessaries at a far cheaper rate than if he dealt in the London markets and shops. The one thing necessary is that the benefits of the scheme should be more widely known. The farmers sell their produce at prices which would open the eyes of the average London householder, and cheaply as the farmers do it they reap nevertheless a substantial profit. And there is absolutely no trouble attending the matter; it is easier for the housewife to send her postal order down to Norfolk or Essex than to go and buy at the stores or the nearest market. The farmers dispatch promptly, and what with our express trains and swift vans the produce is at the housewife's door in a few hours. And perhaps I may suggest that it is a good and essentially patriotic task to support the struggling farmers of eastern England and use home-grown instead of foreign produce. I can not say whether the other railway companies will follow suit, but I believe the scheme will prove to be so beneficial to the public and the companies alike that imitation will surely follow. I may say that the traffic is not particularly remunerative to the Great Eastern at present, nor do I think that, *per se*, it will ever be a highly paying traffic, but it will pay in the long run indirectly by giving a vast encouragement to dairy farming and by spreading a greater prosperity throughout the districts which the Great Eastern serves.

At the half-yearly meeting of the shareholders of this company held in July, 1896, Lord Claude Hamilton, the chairman, announced that the scheme was an assured success; that it was appreciated alike by producer and consumer, and that the company was handling 1.200 boxes per week. The development of the scheme is being watched with interest by the other railway companies and by the wholesale men in London, and it is thought by railroad experts that it may attain very considerable dimensions.

THE SOUTHEASTERN RAILWAY COMPANY.

The Southeastern Railway Company serves the rich and fertile southeastern counties, which have suffered with peculiar severity from agricultural depression, as the lands are chiefly arable and better adapted for the growing of cereals than for grazing. This company has not yet inaugurated any new departure, but has paved the way by calling a conference between its officials and the representative agriculturists along the line. "The fact was recognized," says the report of this meeting, "that in the points at issue the railways and the farmers are practically partners in business, inasmuch as depressed agriculture and diminishing population mean decreased traffic."

7189—No. 12——4

The chairman, the report continues, assured the gathering of the extreme sympathy that the company felt for the farmers, and declared that it was really desirous of doing what it could for them. It would cordially welcome any suggestions that might be made, feeling, as it did, the great advantage of a friendly understanding, and it would be prepared to meet the requirements of the agriculturists frankly, fairly, and generously. Thereupon a large number of gentlemen gave their views on the subject, these views turning mainly on proposals for reductions of rates. It was intimated that at a preliminary conference of the delegates it had been decided to ask for a general reduction of 25 per cent in the rates for agricultural produce; and much was said in favor of reductions for particular articles of traffic, such as fruit, hops, vegetables, milk, cattle, sheep, hay, straw, poultry, etc., complaints being made that in many instances the rates were too high to allow of such commodities being sent to market at all. The paying of higher rates from intermediate stations was also objected to, and through rates for consignments passing over the lines of different companies, a reduction of the minimum in regard to produce in bulk, equality with the foreigner in the proportion of half the same rates for half the same distance, and other similar concessions were asked for, one of the speakers suggesting, amid cheers, that the farmers wanted now to have something more than sympathy, with which they were surfeited.

The speeches delivered were, in the main, thoroughly practical, and at the close Sir George Russell declared once more that the company was most anxious to serve the agriculturists, and sincerely hoped that the proceedings of that day would do something toward increasing their prosperity. He asked the delegates to form a committee of twelve with whom the directors could consult as to details after they had themselves carefully and exhaustively considered the various points that had been advanced.

The report of this gathering is given with a certain degree of fullness, because it is typical of the attitude of the great railroad companies of England toward the farmers. It is a policy of conciliation. It is a policy founded upon the principle, speaking broadly, that there is a community of interests between the patrons of the road and the road itself. No doubt the policy is to some extent forced upon the companies by a powerful public sentiment, as great corporations are notoriously and necessarily somewhat unsympathetic bodies, but the practical recognition of the sentiment, before it has developed into a rooted and dangerous antagonism of spirit, is the part of wisdom.

I have outlined the policy of four of the great railroads of England toward farmers—railways aggregating capital to the amount of $1,400,000,000, and with a total length of about 5,000 miles. Other lines are not behindhand, but it would be repetition to describe in detail their action.

I think on the whole the companies are disappointed at the slowness with which the idea of cooperative or combined shipments is taken up by farmers. The London Times thus comments on the attitude of farmers:

The misfortune is that the English farmer is not easily to be induced to move out of the beaten track which he knows. His love of independence is an obstacle in the way of his falling in with any joint scheme of cooperative distribution, and he is too suspicious or too distrustful of novelty to hope much from any proposed change. There must be a twofold cooperation. The farmers must combine with one another to aggregate their consignments so as to send them in bulk, and the railways must do what they can to make combination easy for them, and must offer a reduction of rates as an inducement to them to enter into it. However great the good will of the railways, and however effective the pressure put upon them by the president of the board of trade, the grievance of the farmers in the matter of carriage charges will remain unremedied as long as they reject the only terms on which a remedy is to be had.

TYPICAL ENGLISH RATES ON AGRICULTURAL PRODUCTS.

LONDON AND NORTHWESTERN RAILWAY RATES BY MERCHANDISE TRAIN (ORDINARY AND EXPRESS) TO LONDON.

[*Explanation of terms.*—C. and D. means including collection and delivery, and such rates are, unless stated to the contrary, applicable to any weight, subject to small-parcels scale contained in pages 129 to 140 of the classification. S. to S. means station to station, and rates noted with this condition do not include collection or delivery. O. R. means owner's risk.]

Rates for grain.

From—	Distance in miles.	S. to S., 2-ton loads (per ton).	S. to S., 4-ton loads (per ton).
		s. d.	*s. d.*
Fenny Stratford	47	7 0 ($1.70)	6 6 ($1.57)
Roade	59	8 9 ($2.13)	7 10 ($1.90)
Buckingham	60	9 2 ($2.23)	7 10 ($1.90)
Brackley	67	9 7 ($2.33)	8 9 ($2.13)
Long Buckby	75	9 7 ($2.33)	8 9 ($2.13)
Coventry	93	10 6 ($2.55)	10 0 ($2.43)

Rates for vegetables.

From—	Distance in miles.	Vegetables in Class C, except cauliflowers, cabbages, celery, rhubarb, and green peas. S. to S., 4-ton loads (per ton).	Mangolds and turnips for cattle feeding, in bulk. S. to S. (per ton).	Remarks.
		s. d.	*s. d.*	
Bedford	62	6 8 ($1.54)	5 6 ($1.34)	4-ton loads.
Blunham	68	6 8 ($1.54)	5 8 ($1.38)	Do.
Potton	74	6 8 ($1.54)	5 8 ($1.38)	Do.
Old North Road	81	7 5 ($1.80)	6 6 ($1.58)	5-ton loads.
Lords Bridge	86	8 9 ($2.13)	7 0 ($1.70)	Do.

Rates for hay.

From—	Distance in miles.	Hay, not machine pressed. Minimum, 30 cwts. per truck. S. to S. (per ton).
		s. d.
Tring	31	9 9 ($2.37)
Bletchley	46	12 8 ($3.08)
Winslow	53	13 10 ($3.36)
Padbury	57	14 7 ($3.54)
Buckingham	60	15 0 ($3.65)
Bicester	65	15 9 ($3.83)
Brackley	67	16 0 ($3.89)

Rates for straw.

From—	Distance in miles.	Straw, not machine pressed. Minimum, 20 cwts. per truck. S. to S. (per ton).
		s. d.
Tring	31	10 0 ($2.43)
Bletchley	46	13 1 ($3.18)
Winslow	53	14 10 ($3.60)
Padbury	57	15 9 ($3.83)
Buckingham	60	15 9 ($3.83)
Bicester	65	16 7 ($4.03)
Brackley	67	16 7 ($4.03)

Rates for fruit (ripe) in Class 2.

From—	Distance in miles.	C. and D. (per ton).
		s. d.
Berkhampstead	27	15 10 ($3.85)
Aylesbury	42	17 6 ($4.25)
Bletchley	46	17 6 ($4.25)
Blunham	68	18 4 ($4.46)

Rates for fresh meat.

From—	Distance in miles.	Rate per ton.	Remarks.
		s. d.	
Peterboro	108	33 9 ($8.20)	C. and D.
Crewe	157	42 0 ($10.21)	Do.
Mold	192	50 0 ($12.16)	Any weight carted in London.
Nannerch	198	50 0 ($12.16)	Do.
		45 0 ($10.95)	1 ton carted in London.
		37 6 ($9.11)	2 tons carted in London.
		32 6 ($7.90)	3 tons carted in London.
Carlisle	298	62 6 ($15.19)	C. and D.
Dumfries	331	75 0 ($18.24)	Do.
		67 6 ($16.41)	C. and D., O. R.
Hawick	344	75 0 ($18.24)	C. and D.
		67 6 ($16.41)	C. and D., O. R.
Aberdeen	539	75 0 ($18.24)	C. and D.
		67 6 ($16.41)	C. and D., O. R.
Keith	592	85 0 ($20.68)	C. and D.
		77 6 ($18.84)	C. and D., O. R.

RATES BY PASSENGER TRAIN.

The rates by passenger train are exclusive of collection and delivery except where otherwise stated.

Rates for milk at owner's risk. *

From—	Distance in miles.	Rate per imperial gallon.	Minimum charge per consignment.
		d.	s. d.
Watford	17	½ (1 cent)	0 6 (12 cents)
Rickmansworth	22	½ (1 cent)	0 6 (12 cents)
Tring	32	¾ (1½ cents)	0 9 (18 cents)
Cheddington	36	¾ (1½ cents)	0 9 (18 cents)
Marston Gate	39	¾ (1½ cents)	0 9 (18 cents)
Leighton Buzzard	40	¾ (1½ cents)	0 9 (18 cents)
Verney Junction	56	1 (2 cents)	1 0 (24 cents)
Claydon	58	1 (2 cents)	1 0 (24 cents)
Islip	72	1 (2 cents)	1 0 (24 cents)
Old North Road	82	1 (2 cents)	1 0 (24 cents)
Hindlow	200	1¼ (2½ cents)	1 3 (30 cents)
Broxton	179	1¼ (2½ cents)	1 3 (30 cents)

*The charges are based upon the actual contents of the cans, and the latter when returned empty are carried free.

Rates for butter, cream, and eggs.

From—	Distance in miles.	Rate per cwt. Minimum 1 cwt.	Remarks.
		s. d.	
Boxmoor	24	1 0 (24 cents)	Butter, cream, and eggs.
Marston Gate	39	1 4 (32 cents)	Cream.
		1 8 (40 cents)	Butter and eggs.
Leighton Buzzard	40	1 4 (32 cents)	Cream.
		1 6 (36 cents)	Butter and eggs.
Armitage	121	2 0 (48 cents)	Cream.
		2 9 (66 cents)	Butter and eggs.
Acton Bridge	173	3 9 (91 cents)	Butter, cream, and eggs.

Rates for fresh meat.

From—	Distance in miles.	Rate per cwt.	Remarks.
		s. d.	
Boxmoor	24	1 0 ($0.24)	Minimum, 56 lbs.
Wolverton	52	2 3 ($0.54)	Do.
		1 9 ($0.42)	Minimum, 5 cwts.
Newtown	195	4 1 ($0.99)	Minimum, 56 lbs.
		3 6 ($0.85)	Minimum, 5 cwts.
Llanidloes	208	4 3 ($1.03)	Minimum. 56 lbs.
		3 6 ($0.85)	Minimum, 5 cwts.
Dundee	471	4 3 ($1.03)	C. and D.; minimum, 1 cwt.
		4 0 ($0.97)	C. and D.; minimum, 1 ton.

Rates for game, poultry, and rabbits (dead).

From—	Distance in miles.	Rate per cwt.	Remarks.
		s. d	
Aylesbury	43	1 6 ($0.36)	Minimum, 56 lbs.
Craven Arms	182	3 9 ($0.91)	Do.
Welshpool	182	3 9 ($0.91)	Do.
		3 3 ($0.79)	Minimum, 5 cwts.
Oswestry	190	3 9 ($0.91)	Minimum, 1 cwt.
Leominster	201	3 9 ($0.91)	Minimum, 56 lbs.
Llangefni	250	4 3 ($1.03)	Do.